PROLO

When I (Ed Close) started writing thi
ago, it was intended to be about how to solve the Rubik's cube and how solution of the three-tiered cube could be used to simulate certain aspects of mathematics, quantum physics, elements of the Periodic Table, cosmology, and the physical, mental and spiritual development of human consciousness. This was essentially done in about 2 years, and about 150 pages. But I thought that the last part of it, concerning the spiritual development of human consciousness, needed to be substantiated and validated with factual events from personal experience, and as I began to write about the relevant events, I realized that many things experienced by Jacqui, my life-partner of more than 40 years, comprised a very important part of the story. The relevant events in her life were entwined with mine, even long before we met in this life. So I asked her if she would tell her part of the story in the book, in her own words. She agreed with some reluctance, and began to write.

As the book evolved, it became clear that, among other things, it was going to be a joint autobiographical sketch of our lives together, focusing mainly on our mission of learning, growing spiritually and helping others. The book shows how the cube can be useful as a physical representation of the nine-dimensional structure of the reality we experience. With three sets of three orthogonal tiers of sub-cubes, rotating with respect to each other, it can simulate our personal experiences of cause and effect and recurring patterns. The patterns observed while solving the cube are analogous, in many ways, to the patterns that arise in our lives. The book offers a framework for analyzing life's challenges and explaining life's mysteries. Our work in this world, however, was not over yet, our story was continuing; so how would we end the book?

As we tried to find a way to finish the book, Jacqui's health began to deteriorate as a result of some issues from the past: she had survived cancer in 1985, and acute kidney failure in 2012. But most people had no inkling of the health problems and the pain she had to endure, because she had the amazing ability to put it all aside and be bright, cheerful and lovingly attentive when interacting with others. She told me more than a year before she passed, that she knew her time remaining in this life was limited, and that God was telling her to start preparing for her transition. Among other things, she began thinking about ways to make things easier for me when she left the body permanently. As the time approached, she began making her final wishes known in detail.

Among other things, she wanted me to finish this book, and she wanted me to continue my research and efforts to get the ideas behind Post-materialist science accepted. She was hospitalized in mid-July 2018 in a lot of pain, but she also experienced several intensely spiritual near-death, out-of-body experiences. She decided to go into hospice December 10th and left the body for the last time on December 15, 2018. In addition to providing a glimpse into the past, explaining why and how we were brought together for a purpose, the book includes what Jacqui wrote about her life before we met, her experiences while in the hospital, and it ends with the documentation of her ascension to a higher dimensional domain, where she was joyously welcomed by those spiritual guides who sent us into this life.

Jacqui as most of us remember her

PRAISES FOR SECRETS OF THE SACRED CUBE, A COSMIC LOVE STORY

"Most people feel that modern physics and higher mathematics appear hopelessly cerebral and utterly remote from human experience and the way we lead our lives. **Secrets of the Sacred Cube** *reverses this pattern. By employing the unscrambling of a Rubik's cube as a metaphor, Edward R. and Jacquelyn A. Close show how the disorder and chaos in our life can be transformed into harmony, beauty, and joy by identifying with fundamental, Primary Consciousness. Secrets of the Sacred Cube is a powerful rejoinder to the morbid scientific materialism of our time. What makes this premise earthy, tangible, and compelling is the interwoven journey of this loving couple — a scientific adventure and a love story rolled into one."*

~~ Larry Dossey, MD, Author: *One Mind: How Our Individual Mind Is Part of a Greater Consciousness and Why It Matters*

Ed, I just finished reading your book and my thoughts are: I love all of it. Much of it is strikingly familiar to me, some of it is new and opens the mind's eye a bit as only sharing thoughts can. Thanks for letting me read it. After reading Chapter IX, I can't say enough. The inspiration is the same, but I remain speechless!

Emotions of divine embrace
Rays of light Christs' shield,
Harmony fills endless space
Truth is Being revealed
Inspired Word of thought
Open doors thought closed

Movements of the symphony
Dance yet being composed!
Thank you for sharing your stories, spiritual perspectives and academic skills.
The Cube Book is inspiring on many levels, many should understand the source of inspiration, and be inspired too!

~~ Your www.friend Patrick Lapham, Sr. Steamfitter LU636

COMMENTS ABOUT THE WORKS OF EDWARD R. CLOSE, PHD, PE, JACQUELYN A. CLOSE, RA, CCI

Abundant prayers and ongoing blessings for the work you and Jacqui have been so dedicated to. Each of you hold a very special place in my heart for all that you have brought and so generously shared with this world. Godspeed.

- Pushkara Sally Ashford

"Our house which was built almost 80 years ago had mold in it. And mold causes all kinds of things, including confusion. And it can make you feel like you've got the flu. Well, **Jacqui Close** *introduced me to something {they call EOB2}, and Jacqui I am eternally grateful to you for learning about this oil, because it does destroy mold. Most recently we had an infestation of mold in our basement, and diffusing [the recommended] oil totally transformed the basement. And now we have that wonderful odor of {the essential} oil. Every household needs to diffuse the oil [recommended by Dr. Close]."*

~~ C. Norman Shealy, MD, PhD, neurosurgeon, psychologist, author, and founding president of the *American Holistic Medical Association*, at the forefront of alternative medicine and alternative health

"It was wonderful hearing Ed present at the ASCI meetings. ... I found myself understanding, appreciating, and celebrating what the two of you have been doing. In fact, as I look over the 40 years of my attending scientific meetings, Ed's presentation is among THE MOST MEMORABLE AND MEANINGFUL of my entire academic (and personal) life".

~~ Dr. Gary E. Schwartz, PhD, Professor of psychology, medicine, neurology, psychiatry, and surgery, University of Arizona, Tucson

"Dr. Close is not only well educated and knowledgeable, but he is a skilled teacher, making complex subjects understandable to the lay person. I have learned a great deal from his books, lectures, and the materials on his website. He is detailed, yet keeps the information interesting."

~~ Dr. Joy Linsley, PhD, Fulbright scholar and educator, Rice University, University of Nottingham, UK, and St. Thomas University, Houston Texas.

"For friends who don't know Edward R. Close, he's the one I'm always quoting from Mold Rx book and completely different title: Transcendental Physics. Genius with great heart and functional intelligence applied to serve the world."

~~ Rev. Lindsay Babich, *Alliance of Divine Love*, Interfaith Minister, Certified Healing Touch Instructor, international speaker and author

*"I would go to my office and be groggy and have a headache, because there was mold present there. Once I got the [recommended essential oil], thanks to **Jacqui Close**, within 30 minutes of turning on the diffuser, I felt better."*

> ~~ Brenda Watson, CNC, renowned natural health and nutrition expert, speaker and author

"Dr. Close has tremendous insight into how to focus the problem, using a methodology that is replaceable, applicable and safe and most of all he does it with grace and ease."

> ~~ Don Clair, PhD, CEO, Clair Caring Center, Greater Atlanta Area, Anthony Robbins Trainer, Drug and Alcohol Treatment Counselor

"Dr. Close is more than well-educated, he is also open to new ideas and exploring them. His forward thinking has helped him to discover things that have never been known to people in our age. I highly recommend reading his publications and learning from his experience any way available to you."

> ~~ Dana Christisen, Complimentary Alternative Health Practitioner and Certified Health Coach

PRAISES FOR *REALITY BEGINS WITH CONSCIOUSNESS* BY VM NEPPE AND ER CLOSE (5TH EDITION, 2014)

"This is the book of books. Close and Neppe have succeeded in articulating a unified theory that explains everything known through human experience and observation, including, not only the data obtained by the five senses, but also the data that comes through mind and feeling. Former scientific thought has considered the material universe to be the total universe with consciousness to be the result of matter. Neppe and Close have shown the reverse. Matter is inseparably linked with consciousness. This is a book you will want to study, absorb, and return to again and again to experience the thrill of understanding how the billions of bits of the universe all fit together as a unified whole, and how we are a participatory part of everything. The author's many years of labor will be appreciated for centuries to come."

~~ David Stewart, PhD, Marble Hill, Missouri, USA. Geophysicist, Theologian, and Author

"Most physical TOEs (theories of everything) fail because they don't explain where the TOE comes from in the first place -- a creative act of consciousness. Most psychological TOEs fail because they don't appreciate the very real physical context in which the psyche struggles to explain itself. Reality Begins with

Consciousness (RBC) avoids these mistakes by taking a TOE's promise of "everything" seriously. This puts RBC in a radical multidisciplinary class by itself, and as a result, understanding it is nontrivial. This should not be surprising, for unadorned Reality as-it-is is vast and hyper-complex, and any TOE that hopes to model that Reality must be equally vast and complex. But for readers who are up to the challenge and are able to stretch their minds in many directions, tackling RBC is well worth it."

~~Dean Radin, PhD, Senior Scientist, Institute of Noetic Sciences, Petaluma CA. Extensively published author, cognitive psychologist, electrical engineer, consciousness researcher.

"Vernon M. Neppe and Edward R. Close have written what is destined to become a classic in the literature on shifting paradigms and worldviews. Drawing from a dozen different disciplines, they have adroitly pointed out the limitations of the Western world's currently accepted model of reality, have spelled out the unfortunate consequences of this model's hegemony, have proposed a paradigm that is not only multidimensional but metadimensional, and have supported it with logic, mathematics, research data, and common sense. The implications that this book has for the betterment of humanity makes for worthwhile, illuminating and enlightening reading which is practical and transformative."

~~Stanley Krippner, PhD, San Francisco, CA, USA. Pioneer, Humanistic Psychology and Consciousness; Saybrook University; extensively published Author and Scientist.

"The Neppe-Close paradigm now provides for a much more coherent way to understand reality. Once introduced, the *actualities of these unifying concepts begin to live. It is quite stunning to observe people speaking from an "already having changed" perspective...a thrilling journey!"*

~~Alan Bachers, PhD, Northampton, MA. Neuroscientist; Director, Neurofeedback Foundation

"Scientific revolutions require both empirical evidence and related coherent explanatory frameworks. The encyclopedic book "Reality Begins with Consciousness" leads in providing the sought after broader scientific unification linking the neurosciences, consciousness, biological, psychological and the physical sciences with math-based logical philosophy and spirituality. Drs Neppe and Close provide a missing broad exploratory paradigm for new scientific ideas that can continue to be researched for many years to come. Whether or not this multidisciplinary model is ultimately viable, the cogent supportive data should encourage scientists to explore seriously the underlying ideas; the models presented go further than other volumes."

~~John Poynton, PhD, London, UK. Consciousness Researcher; Biologist; Past President, SPR. Author.

"Prof. Vernon Neppe and Dr. Edward Close have prepared a much-needed volume that aims to integrate our scientific knowledge into a comprehensive natural-law paradigm. Their work leaves no stone unturned in the quest to re-configure our understanding of science, including those more remote or fringe areas of science such as parapsychology that only a few of our highly respected and honored scientists are brave enough to endorse. This new book by Neppe and Close is a paradigm shift that hails in, if not, beckons for, a kind of scientific overhaul and

shift in thinking that Thomas Kuhn spoke of in his major work "The Structure of Scientific Revolutions."

~~Lance Storm PhD, Adelaide, Australia. Consciousness researcher, Author, Journal Editor, Psychologist, Parapsychologist and Philosopher of Science

"I've just completed the perusal of your impressive work. I feel very excited to have sensed myself, the enthusiasm of discovering so many overlapping fields in our views!... A work that will change mankind's future.
...For the first time in mankind's history, its real nature is scientifically disclosed at the highest charismatic academic level! ... Reading your masterpiece, ... be aware of my deepest reverence for your monumental work!
...A seismic shift in understanding the understanding process itself!
...The beginning of the ultimate disclosure about the nature of an all-encompassing reality.
...A monumental work forcing obsolete preconceptions to crumble.
...The 21st Century's revolutionary paradigm shift."

~~Dr. Adrian Klein, DD, PhD, Israel. Dimensional Biopsychophysicist and Consciousness Researcher. Expert on Theories of Everything, pioneer of the Subquantal Integration Approach;

"This authoritative work in consciousness studies will shape philosophical discourse about mentality and mind. It is a serious and lucid exploration of a most complex topic, suitable for philosophers and cognitive scientists who seek explanatory models that allow one to reach beyond methodological boundaries while at the same time adhering to scholarly rigor. Recognizing consciousness without boundaries and as formative action leads to unexpected conclusions outside any normative space, while

at the same time providing a profound value to the future of humankind. Neppe and Close have masterfully demonstrated that hope is inside and outside Pandora's box."

~~Helmut Wautischer, PhD, San Rafael, USA. Philosopher, Consciousness Researcher and Author. Sonoma State University

"Your book is impressive. It seems astonishing that you could combine deep scientific notions with mysticism. I never knew that such work was possible."

~~Dr. Frank Luger, International (Private) Research Professor, Retired Physician, Mathematical Physicist, Poet, Author, Psychologist, and Chess Grandmaster.

"Broad in scope, multidisciplinary in approach, this monumental work is more than food for thought - it's a feast. If consciousness shapes reality, then you may well adjust your own after reading this book."

~~James Hardenbergh, Seattle, WA, United States. Attorney.

"Without proper scientific evidence or reasoning, consciousness has, for years, been treated as a secondary phenomenon that is somehow derived purely from materialistic origins. There is actually no scientific proof that this is true. I am glad, therefore, that Vernon Neppe and Edward Close have taken on this monumental task of addressing this unwarranted bias in scientific thinking."

~~Kenneth Chan, Singapore, ISPE member and Author.

"Reality Begins with Consciousness is not easy for me, but it is very impressive. It is slowly fascinating me and enlightening. ... The new information has caused great excitement in me. I GET IT! ...at least enough to satisfy my current state of curiosity and need.... I am so impressed with the authors' command of the language. I have to read slow and aloud... to help me digest it. It works well for me. ... By the time I got down to this quote, I was really ready for it! Nice! ..."

~~Carl Lackey, Raleigh, North Carolina, United States.
IT Engineer.

THE AUTHORS

Edward R. Close PhD, PE (Ret.) Diplomate. International Society for Philosophical Enquiry (ISPE), Distinguished Fellow, Exceptional Creative Achievement Organization (ECAO). Dr Close is physicist, mathematician, cosmologist, environmental engineer and Dimensional Biopsychophysicist. *Transcendental Physics* is one of Dr. Close's 8+ books. (www.erclosetphysics.com).

Jacquelyn A. Close, Certified Instructor (Ret.), Center for Aromatherapy Research and Education (CARE) and RA (Registered Aromatherapist) (Ret.) National Association for Holistic Aromatherapy (NAHA), https://www.facebook.com/jacqui.close

Vernon M. Neppe MD, PhD, Fellow Royal Society of South Africa, Diplomate, ISPE, Founder, Pacific Neuropsychiatric Institute, Seattle;(Distinguished Fellow, Distinguished Professor, ECAO. Prof. Neppe is a Behavioral Neurologist, Neuropsychiatrist, Neuroscientist, Psychopharmacologist, Forensic specialist, Psychiatrist, Phenomenologist, Neuroscientist, Epileptologist, Consciousness Researcher, Philosopher, Creativity expert, and Dimensional Biopsychophysicist. Dr. Neppe's 10+ books (http://www.vernonneppe.org/about.php)

OTHER BOOKS BY EDWARD R. CLOSE AND JACQUELYN A. CLOSE

THE BOOK OF ATMA (Evolution of the Soul); Libra Publishers, Inc., Roslyn Heights, New York, 77-87461 (1977)

INFINITE CONTINUITY, A Theory Unifying Relativity and Quantum Physics; Close & Close, Los Alamitos, CA. 1990

TRANSCENDENTAL PHYSICS, Integrating the Search for Truth; First Place winner, Non-Fiction Book Category, Heartland Writers Guild, 1997; First published by Gutenberg-Richter Press, Stewart Publishing, Marble Hill, MO, 1997; toExcel Press, an imprint of iuniverse, Lincoln, NE. 2000

THE BIG CREEK BOOK, History, Folklore and Trail Guide; Paradigm Press, Jackson, MO. 2003

NATURE'S MOLD RX, The Non-Toxic Solution to Toxic Mold; Edward R. Close and Jacquelyn Close, EJC Publications, Jackson, MO. 2007

ANCIENT WISDOM, MODERN SCIENCE, An Integrated Theory of Health; Edward R. Close and Jacquelyn Close, Essential Science Publishing, Salt Lake City, UT. 2010

SPACE, TIME AND CONSCIOUSNESS, Dimensional Linkage; Edward R. Close and Vernon M. Neppe; *Dynamic Journal of Exceptional Creative Achievement 1202:1202; 1101 -1115.* 2012.

REALITY BEGINS WITH CONSCIOUSNESS, A Paradigm Shift that Works; Vernon M. Neppe and Edward R. Close, E-book, available at www.BrainVoyage.com, (5TH EDITION) 2014

IS COMNSCIOUSNESS PRIMARY? Perspectives from Founding Members of the Academy for the Advancement of Postmaterialist Sciences; Editors: Gary E. Schwartz, PhD and Marjorie H. Woolocott, PhD; Chapter 4: Edward Close, "The Mathematical Unification of Space, Time, Energy and Consciousness" In Press, 2018

THE SEARCH FOR CERTAINTY, An Autobiography, unpublished manuscript

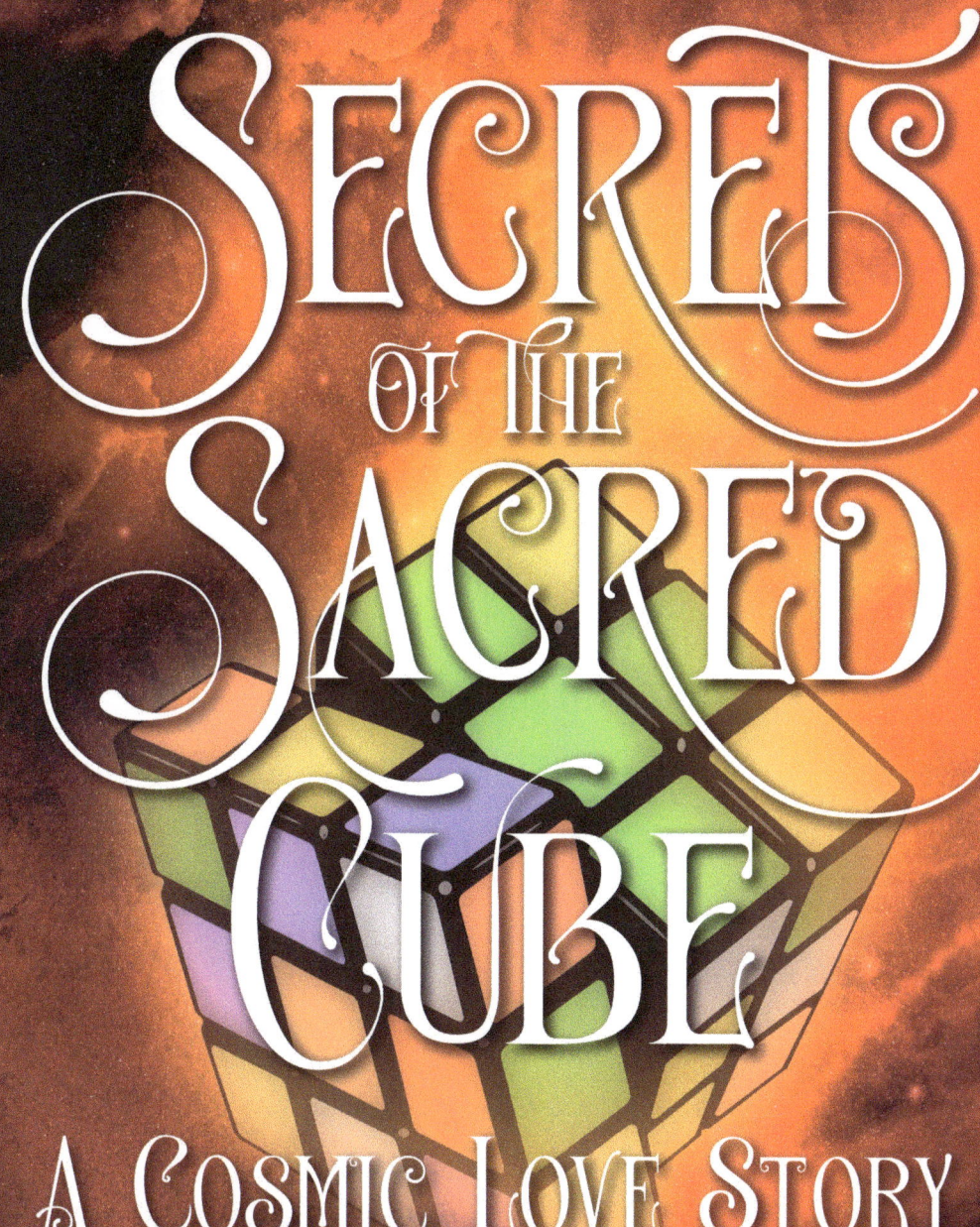

SECRETS OF THE SACRED CUBE
A Cosmic Love Story
All Rights Reserved.
Copyright © 2019 Edward R. Close and Jacquelyn A. Close
v8.0

The opinions expressed in this manuscript are solely the opinions of the author and do not represent the opinions or thoughts of the publisher. The author has represented and warranted full ownership and/or legal right to publish all the materials in this book.

This book may not be reproduced, transmitted, or stored in whole or in part by any means, including graphic, electronic, or mechanical without the express written consent of the publisher except in the case of brief quotations embodied in critical articles and reviews.

Outskirts Press, Inc.
http://www.outskirtspress.com

Paperback ISBN: 978-1-9772-1024-1
Hardback ISBN: 978-1-9772-1025-8

Cover Photo © 2019 Edward Close. All rights reserved - used with permission.

Outskirts Press and the "OP" logo are trademarks belonging to Outskirts Press, Inc.

PRINTED IN THE UNITED STATES OF AMERICA

ACKNOWLEDGEMENTS AND DEDICATIONS

In memoriam: Jacquelyn Ann Hill-Close
1953 - 2018

This book would never have come to be without the loving support of my beloved partner of this life for more than forty years. Parts of this book were written by Jacqui, primarily the parts describing events that happened in her life before we met, and the recent experiences that occurred while she was in the hospital during the period July13 to August 23rd, 2018. But she also inspired much of the rest of the book. She read and made comments that improved the accuracy and readability of certain parts of the book substantially. She was my co-author in more ways than one. If I am ever able to '*soar like an eagle*', to employ the words of one of the songs chosen for the celebration of her ascension, she is truly '*the wind beneath my wings.*'

Honoring my Mother and Father:
Edward T Close and Bernice M Tyndall-Close

Those who know me through my writings, radio talk show interviews, You-tube videos, and presentations at conferences, will not be too surprised to hear me say that I chose these two wonderful people to be my parents before entering upon this life. They loved, raised and nurtured me, telling me that I could be anyone I wanted to be, and do anything I wanted to do. My father was born on January 21st, 1908, My Mother on January 31st, 1917. They were raised in two different valleys about 20 miles apart in the St. Francois Mountains of Southern Missouri. They were poor, hard-working hill folk, whose parents were sons and daughters of immigrants. Typical Americans of their time. My father served in the US Army shortly after the end of World War I, and in the Navy in World War II. My mother was a housewife who dreamed of being a dancer as a young girl, but never had the opportunity to realize her dream. She taught me to read, write, and basic math before I entered kindergarten in St. Louis Missouri in 1941.

TABLE OF CONTENTS

FOREWORD i
PREFACE vi
INTRODUCTION ix
    The Mysteries of Conscious Reality and The Cube ix
    Why the Cube? xi
    Technical Concepts xii
    Life and the Nine Dimensions of Reality xv
    The Purpose of This Book xviii
PART I: KNOWING YOUR CUBE 1
    The Basics 1
    How to Solve the Cube 8
PART II: SOLVING THE MYSTERIES OF QUANTUM REALITY 33
    Reality is Quantized 33
    Simulation of the Combination of Elementary Particles 35
    Combinations of Elementary Particles and Fermat's
    Last Theorem 42
        The True Quantum Unit and The Third Form of Reality ... 43
        The Primary Level of Symmetric Stability – Quarks in
        True Units 49
    The Symmetric Stability of Atoms 58
    The Elements of the Periodic Table 61
    The Cube and the Quantum World 63
    The Origin of Mass 66
    Simulating Atomic, Molecular and Galactic Structures with
    Cubes 75
**PART III: SOLVING THE MYSTERIES OF CONSCIOUSNESS
AND LIFE** 77
    Memory and Intuition 77
    The Elegance of the Three, Six and Nine, and the Cube 82
    The Vehicle of Consciousness 84
    Sacred Dimensionometry and Alignment 85
PART IV: SOLVING THE MYSTERIES OF THE COSMOS 87
    Embedded Dimensional Domains 87
    The Scientific Basis of Infinite 3-D Time 89
    Dark Matter and Dark Energy 100

Modeling the Cosmos 104
PART V: THE ULTIMATE SOLUTION 106
A Comprehensive Approach to Problem Solving 106
When the Cube Comes to Life 107
Conscious Dimensional Mathematics and Mysticism 108
Alignment and Intuition 109
Synchronicity 111
Personal Experiences 111
Belief Versus Knowledge 117
Enlightened Beings 121
Embarking on the Spiritual Quest 127
Experiments in Parapsychology 128
Mathematics, Meditation and Metaphysics 143
Everything is Connected and the Power of Prayer 151
A Couple with a Great Vision: Gary and Mary Young 160
Intellectual Synchronicity: Dr. Vernon Neppe 162
Remarkable Experiences in the Middle East 165

PART VI: PATTERNS OF GOOD AND EVIL 177
Part of a Greater Reality 177
Is Reincarnation Part of Cosmic Recurrence? 178
Conservation of Consciousness 180
The Misinterpretations of the Teachings of Jesus of
Nazareth 182
The Perpetuation of Evil from Justinian I to Adolph Hitler 191
Uncovering the Truth 195
Arguments Against Reincarnation 199
The Origin and Evolution of English Versions of the Bible 202
The Koran, the Holy Book of Islam 205
Evidence for Reincarnation 209
Some Famous People Who have Professed Belief in
Reincarnation 213
Conclusions Regarding the Reincarnation Hypothesis 226
Personal Experiences Suggestive of Past Lives 228

PART VII: CONSCIOUS EXPERIENCES OUTSIDE THE BODY 231
Personal Experiences 231
OBEs and NDEs Experienced by Others 235

PART VIII: THE NATURE OF REAITY, ITS PURPOSE AND GOAL 238
The Goal of Sentient Life 238
Consciousness, Spirituality and the Cube 238
Prayer, Meditation and the Conscious Universe 240

    The Range of Awareness: Minimal to Cosmic 243
    The Future of Science and the Training of Scientists 246

PART IX: A COSMIC LOVE STORY 248
    Struck by Lightning, - and Cosmic Energy! 249
    Six Hundred Miles Away: A Timeless Link 253
    The Meeting in Tampa 257
    Quantum Entangled Lives 272
    Jacqui's Wonderful Life of Service and Devotion 277
    An Important Discovery 292
    Important Networks of Friends 299
    Jacqui's Great Ordeal and Revelations 300

PART X: Cycles of Consciousness and Time 364
    Recent History: the Past 12,500 Years 365
    Have Extraterrestrials visited Planet Earth? 373
    The Way to Enlightenment 379

PART XI: SUMMARY AND CONCLUSION 383
    The Cosmic Love Story 383
    Physics, Spirituality and the Cube 384
    The Message of TDVP, the Cube and the Science
    of the Future 386
    Ode to Joy Now Gone 388

NOTES 389

REFERENCES 393

INDEX 399

FOREWORD

It is a singular honor and a privilege to write the Foreword to this book on 'the magic cube' by Dr. Edward R. Close, physicist and extraordinarily creative mathematician -- possibly the most widely respected thinker in the field of mathematics alive today -- and by his late wife, Jacquelyn Close. Their book, *Secrets of the Sacred Cube: A Cosmic Love Story*, is written from a very interesting and unique vantage point.

Jacqui and Ed are engaging writers, and the book's message is clear and available to any reader. It is a message of hope and bright future for mankind. This book certainly shows how one can more easily provide a solution to the Rubik's cube. When Ed first told me he was writing about this subject, I thought that this was the major focus of his book. However, this is just one relatively small component of a book that is mainly autobiographical, and biographical in regard to describing one's spouse. And so we are able to appreciate the love between Edward and Jacquelyn and trace it over more than 40 years. It is a love that has implications not only for their present life but also thinly veiled within previous incarnations.

I have known Dr. Close for a number of years, and I came to know Jacqui as well because of the special Ed-Jacqui relationship as partners in life. Both have been extremely remarkable people, with Jacqui particularly developing the whole area of the use of essential oils, and recognizing the value of the herbal components. Their deep, complimentary and loving relationship makes them as a couple a very rare one. Their story is outlined in brief vignettes in this book, ostensibly from many lives. It is fascinating, revealing patterns of a Grand Plan, replete with meaning, that involves not only their individual missions, but the purpose and destiny of humankind. I have been fortunate to be just a small part of their destiny.

I myself have spent a rewarding career in medicine, psychiatry, and behavioral neurology and the neurosciences. I have studied the human psyche and the mental, physical, and spiritual characteristics of those with exceptionally high intelligence and creativity. Dr. Edward Close is such a truly exceptional individual. I've worked closely with Dr. Close in multidisciplinary research for more than a decade on a model that we call the 'Triadic Dimensional Distinction Vortical Paradigm' -- or TDVP. It is groundbreaking in that it examines the nine *quantized*, discrete dimensions that are enveloped by a continuous finite. This sometimes has to be modeled and the Rubik's cube somewhat surprisingly fits the model allowing visualization of multiple dimensions.

The Rubik's cube solution is only one component of this book: A major feature is the way Ed shows how this cube and such geometric forms can be applied. This is particularly so in our own Neppe-Close work, in the so-called 'dimensionometry' of TDVP, which is effectively multi-dimensional geometry across 9 dimensions. I remember the amazement of a prominent consciousness researcher when he discovered that Ed was applying the Rubik's cube so one could more easily visualize a 9-dimensional model of reality. And each component -- square, block, cube -- of the Rubik's cube is different, reflecting possibly mimicking the 3 dimensions of space, 3 of time, and the 3 of consciousness, that we have theorized, and for which there is significant mathematical support. Those three components -- of consciousness, space, and time -- are hypothesized; as opposed to a 9-dimensional matrix which is demonstrable mathematically.

Perhaps the most amazing component of this book is not the detailed mathematics and complex concepts, but that they are so accessible. Each component of the book is stand-alone, so one can pick up chapters and read them as real life stories related by both Jacqui and by Ed, with the message being clear and available to others -- to any reader, in fact. The message is one of hope and a bright future for mankind, linked up with the *mental virtue*, with higher consciousness, with the cosmic God, and an awareness of ascensions into eastern philosophies and multi-dimensional time. It's a move away from the negativity of one's essence being trapped in difficulties, and an emergence from the dark ages of materialism, striving for enlightenment.

There are links and resonations of this book with Kabbalah, with Jainism, and with many other venerable metaphysical disciplines; something Dr. Close recognized early on in his *The Book of Atma*. By applying the cube to model the quantum, our micro and macro world, and the cosmic reality, Dr. Close has demonstrated cogently how everything is interconnected, that there is a unified monism and a holistic unity of all things.

It helps to have a profound understanding of physics and mathematics, as he does, and to be able to reflect the logical structure of reality on display, not only in the detailed analysis of the Rubik's Cube but in its use as a simulation device -- and along with all this, the development of a primary quantum calculus designed to reunite the analytical process of theoretical physics with a metaphysical basis of the perennial philosophies of Leibniz and Huxley.

This book should be read with an open mind to understanding components of the detailed research and the meticulous presentation of concepts like quantum equivalence, gimmel, cyclic time, and reincarnation. All of these will give you, the reader, something to think about, regardless of your own religious and cultural background.

I regard Dr. Close as like a brother. We have spent many years trying to understand and explain to others and to the world the new discoveries and the implications of a 9-dimensional plus reality: 9 quantized dimensions with little discrete components -- at the quantum level, the macro level, and the cosmic level: these are all unified into one and enveloped by an infinite continuity. This has been our song to sing, and Dr. Close has been an essential and important component of the singing.

This is not just another message, another book written biographically, a love story, a cosmic romance along with a component of discussing a mundane project with the Rubik's cube. This is far, far greater, reflecting mental virtue and development at both the eastern philosophical level -- with what Dr. Edward Close regards as the *Dwarpara Yuga* (Vedic time) -- and movement away from the lingering zeitgeist of the *Kali Yuga*, a cycle during which many souls are trapped in finite relativity. The emergence from materialism into the age of awareness, consciousness and enlightenment, is something extremely important.

These philosophies resonate with much of my thinking. But the bottom line is not the philosophy; the bottom line is the fact that we have been able to develop methods in line with this approach that can be proven scientifically.

I cannot adequately endorse this work just by these descriptions, because it mixes the esoteric, the unique, and the highly relevant. Indeed, despite the separate components, it can be seen as a unitary autobiography of a remarkable mind.

Professor Vernon M Neppe MD, PhD, Fellow Royal Society (SAf), Distinguished Fellow APA, MMed (Psych), DPsM, FFPsych, DABPN, DSPE, DCPP (ECA)

Director, Pacific Neuropsychiatric Institute, Seattle, WA, USA

Executive Director and Distinguished Professor, The Exceptional Creative Achievement Organization;

Dimensional Biopsychophysicist and Consciousness Researcher

Director, Human Performance Enhancement Division (HPED) of the World Institute of Scientific Exploration (WISE)

Whiting Memorial Prize Recipient (2016) with Edward Close

America's Top Doctors (2001-2018) (current listings under 'Behavioral Neurology and Neuropsychiatry, Psychopharmacology, Psychiatry, Forensic Psychiatry / Neuropsychiatry: peer-reviewed Castle Connolly)
Editor, Dynamic International Journal of Exceptional Creative Achievement

Editor-in-chief, Journal of Psychology and Clinical Psychiatry

Author of 11 books including
Cry the Beloved Mind: A Voyage of Hope
Reality Begins with Consciousness: A paradigm shift that works (with Edward Close)

PREFACE

I purchased my first Rubik's cube in 1980 or '81, just after it came onto the market in the US as a three-dimensional puzzle and toy. I took my brand-new cube with me to the Middle East in 1981, where I was employed as an environmental planner and engineer during the construction of the Industrial Cities of Jubail and Yanbu, at the east and west ends of the trans-Arabian pipeline.

During the month of Ramadan that year, I was in Yanbu, on the Red Sea about half way between the Gulf of Aden and the Holy Land, where we were building a deep-water port, refineries and government and civilian housing at the west end of the Trans-Arabia Pipeline. As anyone who has lived in the Arab world knows, devout Muslims do not work, drink or eat during the daylight hours of Ramadan. As a result, most non-Muslims working in Muslim countries leave before the start of Ramadan and go home or elsewhere during the Holy month because little can be done on projects that require the involvement of local people. I hadn't been in Saudi Arabia long enough to have my family join me, or earn a vacation, so I had some time on my hands, and I used some of that time to solve my Rubik's cube.

There were no books, websites or "cuber" groups as there are today, so I was on my own. It took me about two weeks to work out a way to solve the cube, i.e. restore a mixed-up cube to its original red, orange, white, yellow, green and blue side perfection. No one told me that solving the cube was believed to be virtually impossible without instructions, and that it took the creator of the cube, Erno Rubik, a month to solve it himself.

Unhampered by such discouraging info, I saw the cube as just another challenging puzzle to solve - and solved it.

During the years 1986 to 2000, I spent a lot of time researching the relationships between quantum physics, relativity and consciousness, subjects that had been of great interest to me during my college and university years while I was studying physics, mathematics and environmental engineering, and practicing time-honored consciousness-expanding techniques. In 1977, I published "The Book of Atma", a book on consciousness and spiritual evolution. In 1990 I published my second book titled "Infinite Continuity", introducing and applying a new form of mathematical logic called the Calculus of Distinctions, and in 1997 I published "Transcendental Physics", with proof of the existence of a non-quantum receptor, implying that consciousness, not matter, is the source of objective phenomena.

In addition to those years of research and publishing, after being accepted into membership of the International Society for Philosophical Enquiry (ISPE) in 2006, I teamed up with Dr. Vernon M. Neppe, MD, PhD, a prominent neuroscientist and member of the ISPE who had developed ideas about the nature of reality similar to mine. Together, we have developed the Triadic Dimensional Vortical Paradigm (TDVP), a shift to a consciousness-based reality model. We co-authored a number of papers and articles together, and in 2011, published the book "Reality Begins with Consciousness".

This book, using the cube as a visualization aid, draws on many findings of our research over the years, and presents, what is in effect, a functional "theory of everything" that can be understood by anyone. The book also includes accounts of some relevant personal experiences that relate to the application of the theory to real life. My hope is that reading this book will stimulate some thinking that will be beneficial to the reader. It provides a way of obtaining useful information from the geometry of the Rubik's cube and the TDVP consciousness-based model of reality for application in everyday life.

Edward R. Close, April 22, 2017

INTRODUCTION

The Mysteries of Conscious Reality and The Cube

This book is not just about solving the Rubik's Cube®. It is much more than that. It is about solving the important problems of life. I was pleasantly surprised when I discovered that the cube could be used to model many different aspects of reality, and it was even more surprising to learn *why* that is true. In this book, I shall attempt to show that reality exists as a series of worlds within worlds analogous to the cubes within cubes of the Rubik's cube®.

I will attempt to explain how you can master the cube, and why by doing so you will open the door to a much deeper understanding of reality. I will present a brief history of the patterns of good and evil in the world, illustrating how the simple truth about the basic spirituality of life is easily overlooked, especially if it has been obscured by the misrepresentations of authoritarian institutions for the purpose of the political control of people, power and wealth. I will briefly recount the history of some of the major belief systems in the world, and I will also share some personal experiences that involved synchronicities, intuitions and alignments, reflecting the structure of reality, and reveal how these patterns are reflected in the geometrical structure of the cube, and how this knowledge can help you in your personal journey through life. Sound like heavy stuff? Maybe, but I'm hoping that you will find that learning about it in this book is a lot of fun as well.

The Rubik's Cube® has 9 rotatable planes or tiers, 3 in each of 3 orthogonal dimensions. This book shows how a simple cube can be used to model everything from electrons and atoms to solar systems and galaxies. Solving the cube even mirrors the development of different stages of physical, mental and spiritual health, and teaches us how to develop intuition, focus and concentration, - valuable skills that are needed in order for one to solve the problems of life. In the process, you may also learn how to 'solve' the cube.

This book is divided into eleven parts:

I. **KNOWING YOUR CUBE**

II. **SOLVING THE MYSTERIES OF QUANTUM REALITY**

III. **SOLVING THE MYSTERIES OF CONSCIOUSNESS AND LIFE**

IV. **SOLVING THE MYSTERIES OF THE COSMOS**

V. **THE ULTIMATE SOLUTION**

VI. **PATTERNS OF GOOD AND EVIL**

VII. **CONSCIOUSNESS EXPERIENCES OUTSIDE THE BODY**

VIII. **THE NATURE OF REALITY, ITS PURPOSE AND GOAL**

IX. **A COSMIC LOVE STORY**

X. **CYCLES OF CONSCIOUSNESS AND TIME**

XI. **SUMMARY AND CONCLUSION**

Each of these parts is designed to stand alone, but three threads of **alignment**, **intuition** and **synchronicity**, run throughout the book, and connect them all. I'm hoping that you will find reading this book to be an interesting mental, physical and spiritual adventure involving the cube, geometry, logic, spirituality, science and life. This book summarizes a philosophy of life developed during the eighty-plus years I've spent on this planet as a human being named Edward Roy Close. Any journey, even the journey of a lifetime, begins with a single step, and the adventure of reading a book begins with the reading of the first page. My gratitude and blessings to all who choose to read on.

Why the Cube?

Physically, the Rubik's Cube® is simply a small cube, about 6.5 centimeters (a little over 2 inches) on each side, with each of its six faces divided into nine squares of equal size, totaling 6x9 = 54 exposed squares. A brand-new official Rubik's Cube® fresh from the manufacturer has six monochromatic faces: One face is red, one face is blue, one face is yellow, one face is green, one face is white, and one face is orange. It looks as if it were made by slicing a two-inch cube into 27 smaller cubes of equal size. It turns out, as we shall see in Part I of this book, that this mental image of the cube, like many conceptualizations based on our sense perceptions of things in the physical world, is an illusion. However, in the cube there are nine planes or tiers, appearing to consist of nine small cubes each, that can be rotated either clockwise or counterclockwise, independent of the other eight tiers. This means that there are exactly 9x2 = 18 separate 90-degree rotations possible from any given configuration of the cube at any time.

This book is intended to be much more than just a way to learn to solve the cube. The purpose of the book is to provide the reader with a practical tool that will help in the development of the focus, memory skills and intuition needed to solve practical down-to-earth problems. Analogies relating efforts to solve the cube with real-life problem solving are illustrated with brief discussions of events in the lives of the authors, making the book semi-autobiographical. Our personal experiences illustrating important points are brought up as appropriate in the course of discussing the use of the cube for problem solving. But first, some technical details will be helpful for the reader who wants to understand the cube. The technical details touched on in this section are intended to help the reader understand why and how use of the cube works as a tool to solve problems, but an in-depth understanding of the technical concepts is not necessary for you to begin using the lessons of the cube to solve real problems.

The reader may proceed as he or she chooses: *You can skip over to sections you find more interesting, if you wish. You can always come back and read the technical sections later if you want to. The object is to acquire a new way of thinking and some new tools to help you cope with life. But, don't feel obliged to absorb every word. I want you to enjoy our time together between the covers of this little book.*

Technical Concepts

Because of the nine orthogonal-plane, rotational capability of the cube, it can be used as a physical model to simulate the nine-dimensional reality indicated by the mathematics of the Neppe-Close model of reality, illustrating the dynamic rotational relationships between mass, energy and consciousness at the quantum, human and cosmological levels. The nine planes of rotation can be used to represent nine different dimensions of finite reality. There are three operational principles linking these modeling applications. I call them the three PSRs:

1. The Principle of Symbolic Representation

2. The Principle of Symmetric Resonance

3. The Principle of Synchronous Relativity

Symbolic Representation:

All phenomena are finite representations of Reality that can occur in three forms which are mathematically describable using three types of symbolic variables:

1. Variables of Extent: describing three sets of three dimensions

2. Variables of Content: mass, energy and consciousness

3. Variables of Intent, Impact and Information: existential, conceptual and imaginary

For brevity, I will refer to the Rubik's cube® hereafter simply as "the cube" except in circumstances where the full designation is needed for clarity. The sub-cube pieces and various aspects and configurations can be employed to symbolically represent real phenomena of mass, energy and consciousness.

Symmetric Resonance

There are at least three types of symmetric resonance that can be modeled with a rotating cube:

1. Mass: elementary particles forming the stable physical structures of the universe: Up-quarks, down-quarks and electrons

2. Energy: light, electricity and magnetism

3. Consciousness: individual, group and Cosmic

I'll be mentioning the mathematical logic of the Calculus of Dimensional Distinctions (CoDD), a calculus that I developed over a period of about forty years. It is not necessary to master all of the logic and mathematical operations of the CODD to understand the message of this book, but the CoDD is a simple mathematical system that is logically prior to conventional mathematics, and application of the CoDD to symbolic logic, relativity and quantum physics yields answers to puzzles unresolved by mainstream science using conventional mathematics. With applications of the CoDD, we have found that the nine finite dimensions of extent are embedded in an infinite primary substrate of Pure Consciousness. This allows all phenomena to resonate as parts of the infinite while maintaining a finite symmetric balance in relation to the whole, the totality of reality. Symbolic representation of the interaction of the three forms of resonance evokes the Principle of Synchronous Relativity.

Synchronous Relativity

The symmetric resonance of reality requires synchronous interaction between phenomena across all dimensions. Asymmetries present in any observational distinction drawn by an individualized conscious entity in less than nine-dimensions are balanced by complementary asymmetries in other dimensions of the holistic nine-finite-dimensional domain. This balance and parity are reflected in the symmetry and parity of the cube. The manifestation of reality as it is observed by any individualized conscious entity is incomplete in the dimensional framework available to that individual. Because our physical senses are severely limited, major underlying aspects of reality are hidden from us, and, like the assumed internal structure of the Rubik's cube, some of our assumptions about physical, mental and spiritual reality are simply convenient illusions. When viewed in the context of higher dimensional domains, new relationships are revealed.

Implications

The mathematical logic of the holistic synchronous nature of Reality requires that time and consciousness, like space, are three dimensional. The nine orthogonally movable planes of the cube provide functional representations of the inter-related nine dimensions of cosmic reality, quantum reality and spiritual reality.

Parts of this book are based on lectures and workshops presented from 2011 until 2017 using the Rubik's cube as an aid to visualization of cosmic, quantum and spiritual reality.

Life and the Nine Dimensions of Reality

"Ask, and it will be given to you; seek, and you will find; knock, and it will be opened to you." - Matthew 7:7

Can this Biblical promise be taken seriously? Is there an answer to every question we can think of to ask? Is everything we don't currently know about the nature of reality hidden from us in such a way that, if we can find the right door and knock, it will be revealed to us? Almost every theoretical physicist since Einstein has talked about a "theory of everything". In the 1980s Stephen Hawking predicted that we would have the theory of everything by the year 2000. Of course, that was wishful thinking on his part, and it didn't happen. Can there ever be a theory of everything? Maybe. But the theory of everything that physicists talk about presupposes that there is nothing to reality but matter and energy interacting in space and time. What if there is something more?

In fact, we now know that there is.

Physicist John A. Wheeler, discussing the delayed-choice experiment, indicated why consciousness plays a fundamental role in reality when he said "No elementary phenomenon is a phenomenon until it is a recorded phenomenon." Based on repeatable and verifiable evidence and rigorous mathematical proof, in the Triadic Dimensional Vortical Paradigm (TDVP), developed over the past ten years by Dr. Vernon Neppe and this author, we have demonstrated that there is indeed something more. Furthermore, matter, energy, space and time, as we experience them through the limited physical senses, are not what they seem to be. As Albert Einstein once said: "Reality is merely an illusion, albeit a very persistent one." It is not that reality is willfully hiding itself from us; it is all there for us to find, if we decide we want to know more, and can change our focus from the purely physical, which, while it is fascinating short term, is ultimately a dead end.

The totality of reality, including but not limited to the physical universe, has logical structure, purpose and meaning. And that structure is mirrored by what we know as hyper-dimensional geometry. Actually, it would be more appropriate to call it *dimension*ometry, because the suffix geo- relates it to planet Earth, and the logical structure of reality goes beyond the structure of this small planet; it is galactic and cosmic.

We call our knowledge of the nature of reality *science*, but our current understanding of the nature of reality is still very rudimentary because it is based on indirect evidence and incomplete information. This information is contained in the data conveyed to the imaging function of consciousness in the brain through the physical senses, which only convey a very small fraction of the spectrum of reality into our awareness, and our conceptions of it can be expressed using symbols and models. A multi-dimensional cube with articulated orthogonally rotating planes, like the Rubik's cube, is unusually well suited to the modelling of the physical, logical and spiritual framework of reality.

An Important Discovery

The first glimpse of the potential utility of the cube for explaining the dimensionometric aspects of reality came when I found that the "intrinsic ½ spin" of fermions, the building blocks of the universe, could be visualized and simulated using the cube; and that led to an explanation! I discovered that the simulation provides a definitive demonstration that the intrinsic spin of fermions is simply the natural result of objects spinning simultaneously in three, six or nine dimensions. This was an exciting discovery, but when I found that electrons, quarks, protons, and neutrons could also be modeled using the cube, I knew I was on to something.

Imagine my excitement when I discovered that the cube could be used to simulate not only quantum physics, but also relativistic cosmology, and even spiritual evolution!

The Purpose of This Book

The purpose of this book is to provide the reader with a way to model the three most important aspects of reality: physical, mental and spiritual. Physical reality can be modeled on the quantum-, macro- and cosmological levels. The mental world can be modeled as subconscious, conscious and super-conscious states; and the spiritual world can be modeled on the primary, individual and Cosmic levels. **The goal of this book is to show that these nine levels and states of reality are interdependent, and that, when integrated, form one elegant, consistent whole.**

The intended purpose of this book is to demonstrate how and why the cube may be used to simulate complex physical, mental and spiritual systems, and perhaps even model a functional "theory of everything". This is possible because the cube is designed in a way that it can simulate an internally consistent dimensionometrical system of nine interacting rotating tiers, each ostensibly consisting of nine smaller cubes. Science has shown us, especially with the four-dimensional space-time model of Einstein and Minkowski, the five-dimensional models of Kaluza, Klein, Nordstrom, and Wolfgang Pauli's six-dimensional model, that the physical universe consists of matter, energy and spacetime, inter-related by an invariant mathematical dimensionometry. We now know that consciousness, in its individual, group and cosmic manifestations, interacts with this dimensionometry, and gives purpose and meaning to the physical universe.

The work of Neppe and Close, (Neppe and Close, 2011) extending the multi-dimensional approach to a model of nine finite dimensions embedded in a transfinite domain, has revealed the involvement of a third form of reality that is not detectable as mass or energy. The third form, which we've called *gimmel*, is a non-physical (without mass or energy) component of reality without which there would be no stable physical universe. The conceptual model, called the Triadic Dimensional Vortical Paradigm (TDVP) has been developed using precise experimental data from the Large Hadron Collider and Mathematical theorems, verified by producing explanations of several puzzles that have baffled mainstream scientists for decades.

This book presents some of the scientifically validated applications of the TDVP model, as well as some controversial projections beyond what we have proved. Such projections, especially those dealing with spiritual concepts like the existence of consciousness independent of the physical body before birth and after death, divine origin and reincarnation, must be considered as speculative from a scientific point of view. Data from scientific studies and some circumstantial evidence are provided, including personal experiences so that the reader may decide for him/herself whether these projections are valid or not.

Even though the main purpose of this book is not to teach you how to solve the cube, that may be an enjoyable side effect. If you want, using the methods of solution presented in this book, with focus and practice, you can learn to solve any scrambled cube in as little as one minute or less.

To understand how something as simple as a cube can be used to simulate things as complex and different as particle physics, cosmology and spiritual reality, it is necessary to realize that all plausible finite patterns and forms are reflections of the logical structure of reality, and that the structure of finite reality is pure invariant symmetric dimensionometry in more than three dimensions, which is the mathematical basis of the Calculus of Dimensional Distinctions and the Triadic Dimensional Vortical Paradigm (TDVP), the Close-Neppe model of reality.

As we developed our model, Dr. Neppe and I found that the underlying mathematical dimensionometries of physical, mental and spiritual Reality are very similar, if not cogently identical, conforming to the saying "As above, so below". This is why so many things can be simulated with the cube. **Once this pervasive logical dimensionometrical pattern is understood, we can see that a completely scrambled cube is analogous to an incoherent physical, mental or spiritual state, and a "solved' cube, with each of the six sides monochromatic, symbolizes a symmetrically balanced, aligned and coherent state of being.**

Note: Vortical is the adjective form of the noun vortex.

Contemplating the Cube

PART I: KNOWING YOUR CUBE

The Basics

Let's start by becoming familiar with the cube. Just looking at a Rubik-type cube, it appears to consist of 27 small cubes that are identical miniature versions of the larger cube, positioned together, 3 by 3 by 3, to form the larger cube. It also appears that by twisting and turning the cube, any one of the smaller cubes could be moved from its current positions to any other location in the larger cube. **None of this, however, is true**.

What looks like 27 small cubes with the same six different colors on each of their six sides, stacked together to form the larger cube, *is, in fact, not that, at all*. None of the small cubes are colored on all six sides, none of them are actually cubes, and none of them can be moved to all of the other locations by any amount of twisting and turning whatsoever.

Already we see that the Rubik's Cube is like reality: What it appears to be is not what it really is. In that sense, like life, it is an illusion, and its solution may even seem magical.

We can see what the "small cubes" actually are by taking the cube apart. You will find that none of what you thought of as smaller cubes are actually cubes. The cube consists of three different types of pieces, with one, two or three, square, colored sides, connected by a three-dimensional six-directional axel at the center. The different types of pieces may be identified by how many square sides they have exposed on the surface of the cube and what color they are.

We see that the part in the center of the cube, around which all the other pieces rotate, is a tri-axel connected to six, square single-color faces located in the center of each face of the cube, but their connection is never exposed, unless the cube is broken apart. The six square faces connected by this center piece and located in the center of each face of the cube, never move to any other location relative to each other. Because they never move relative to each other, but only rotate in place, they can be identified by color and used as handy orientation points: six center squares, six different colors. In addition to the tri-axel center piece, there are twelve pieces that have only two colored faces. They are the "edge" pieces, one on each of the edges of the cube. The third type of pieces comprise the eight "corner" pieces that have three colored sides exposed. There are no sub-cube pieces with four or five colored faces, and only one with six, the center with extensions to the six faces, but of course, it is not a cube.

If the cube actually consisted of 27 small moveable cubes, placed in a 3x3x3 block, as seems to be the case at first glance, with each side of each small cube one of six different colors, then the number of different configurations possible would be beyond astronomical: There would be 6x27 = 162 cube faces, 27 of each color, e.g., red, orange, white, yellow, blue and green. The number of faces visible in any specific configuration would be 54, the number of squares visible, occupying the positions available on the outside of a 3x3x3 cube, and the number of possible locations for a given-color face would be 60/6 = 10. This means that the number of different configurations of the small cubes possible would be equal to n! times P^n, where n is the number of faces, n! is the number of permutations of n faces, and P is the number of possible locations of each mini-cube of each of the six colors.

The notation n! is called "n factorial" and is equal to 1x2x3x...xn. Thus n!xP^n = 60 factorial times 10 to the 60th power, which is an unbelievably huge number: (1x2x3x4x5x6x7x8x9x10x11x12x 13x14x15x16x17x18x19x20x21x22x23x24x25x26x27x28x 29x30x31x32x33x34x35x36x37x38x39x40x41x42x43x44x 45x46x47x48x49x50x51x52x53x54x55x56x57x58x59x60) x10^{54}. If you try to carry out these multiplications, the number of digits of the answer will soon exceed the capacity of any computer you might have on hand. This is a truly astronomical number, more than all the stars in the visible universe, more than all the grains of sand on all of the beaches on the planet and all the atoms making up the cells of the bodies of all of the living organisms on Planet Earth, - all added together. This is a good example of how something very simple can easily become very complex very quickly.

The good news, however, is that this huge number of different configurations is irrelevant to solving the cube, because, as we saw above, the Rubik's Cube *does not* consist of 27 cubes with 27x6 = 162 colored faces. Instead, it consists of a number of odd-shaped parts, 8 with three colored faces, 12 with two colored faces, and one with six colored faces, totaling 21 pieces with a total of only 54 colored faces. This is good news, because the number of possible configurations is much smaller than it would be if there were actually 27 small cubes with colors on all sides that we could move independently. We can recalculate the number of possible configurations of the Rubik's Cube that we may be confronted with as follows:

If the movement of the pieces were not constrained by their being part of the Rubik's cube mechanism, the total number of possible configurations would be equal to the number of configurations possible for each of the types multiplied together, i.e., the number of possible configurations for the single-faced pieces times the number of possible configurations for pieces with two colored faces times the number of configurations possible for pieces with three colored faces. In mathematical shorthand that is: $N = (1!x6^1)x(8!x3^8)x(12!x2^{12})$. But since the three types of pieces are inter-related by being attached to the rotational center of the cube, and our object is to find a way to solve the cube, we are not really all that interested in looking at every single one of the possible configurations. That's a good thing, because it would take far longer than your entire lifetime to do that. A computer can do it in less time because it can carry out a great number of operations in a nanosecond. But a computer has no conscious awareness of what it is doing. It can only do what it is programmed to do by a conscious being. You and I, on the other hand, are capable of thinking about the situation from various angles and points of view existing *outside* of any dimensional framework that might be programmed into a computer.

Isn't it good to know that you are actually smarter than your laptop?

When we begin to execute rotations to change the configuration of the individual face pieces of the cube, to mix it up, and/or find a way to solve a scrambled cube, we can orient the cube in only one of six different ways by choosing one of the six faces as the reference face, and the color of the center square of that face as the reference color. When I first solved the cube, I chose red as the reference color, and I generally stick with that to avoid confusion, – and believe me, confusion is a very real possibility when one is trying to solve the cube! Because the cube is symmetric, there are six ways to orient the cube, so there will be six versions of each set of configurations leading to a solution. Each of the six types of pieces will have the same configurations, so we can eliminate 5 out of 6 of them because they are geometrically the same, and therefore redundant.

Starting with the pieces that have only one colored face, we can see that all six combinations given by (1!x6) = 6, have the same geometrical configuration for each color. So, we only have to consider the configuration of one color, the one we choose as our reference. Therefore, for the pieces showing one-side, we can divide the number of configurations by 6, the number of colors, and instead of ($1!x6^1$) = 6 configurations, we need only consider ($1!x6^1$)/6 = 1 configuration. This means that we can divide the total number of configurations by 6.

Visualizations for the two-face piece type and three-face piece type are a little more complicated, but the redundancy for each type can be determined by dividing the number of possible combinations by the number of geometrically equivalent configurations for that type.

The number of geometrically equivalent configurations for the three-face pieces is the number of colors minus the number of sides = 6 − 3 = 3, and the number of geometrically equivalent configurations for the two-face pieces is 6 − 2 = 4. So, we eliminate the redundancy in the possible combinations for each by dividing by the number of geometrically equivalent configurations; so, for N, the number of different configurations that we must consider, we have:

N = [($1!x6^1$)/6]x[($8!x3^8$)/3]x[($12!x2^{12}$)/4] =[($8!x3^8$) x($12!x2^{12}$)]/12

Carrying out these multiplications and divisions, we get the number of non-redundant configurations to be: **N = 43,252,003,274,489,856,000**. This is still a very huge number of different configurations; more than 5,000 times the estimated total number of grains of sand on all the beaches on the planet! Given a scrambled cube, the configuration you are confronted with could be any one of these 43 quintillion-plus configurations. Of course, if it happens to be the one where all six sides are monochromatic, the cube is solved. But the chance of that happening by accident is one in forty-three quintillion, two hundred and fifty-two quadrillion, three trillion, two hundred and seventy-four billion, four hundred and eighty-nine million, eight hundred and fifty-six thousand, and that's equal to the tiny decimal fraction, 0.0000000000000000000231, or 2.31×10^{-20}.

There will be some other configurations that are not far from total monochromatic coherency, from which a solution might be achieved relatively quickly by an intelligent person, but all of those added together would still be a very small number relative to the total number of possible configurations, The fact that it has been proved that there are multiple paths to a solved cube from any given scrambled configuration might also be encouraging, but even if there are a million different paths to solution from any given scrambled configuration you happen to be confronted with, the chances of finding one of them by random trial and error, is 1,000,000 divided by 43,252,003,274,489,856,000 = 0.0000000000000231 = 2.31×10^{-14}. If you tried one path every minute, it would take you more than a trillion years to try them all. This means that, *on average*, it would take 500 billion years to solve a scrambled cube. That's more than 35 times the age of the universe estimated by the big bang theory! This is why some people say solving the cube without some help is virtually impossible. However, ...

Looking at the "good news – bad news" information above about the chances of solving the cube, reminds me of the comedy routine of 'Lonzo and Oscar heard on country music radio when I was a teenager, more than sixty years ago. It went something like this:

One of them says: "Did you hear about Caleb? He's in the hospital"!

No! What happened?

He fell off the roof of his barn.

Oh, that's bad!

Well, not so bad, there was a haystack beside the barn,

Oh, that's good!

No, not so good. There was a pitchfork in the haystack.

Oh, that's bad!

Not so bad, he missed the pitchfork.

Oh, that's good!

No, not so good, because he missed the haystack too!

How to Solve the Cube

The previous section made it clear that solving a thoroughly scrambled 3x3x3 cube by accident is less likely than winning the lottery every day for a year, - not impossible, but it might as well be! But we see hints that there may be clues in the patterns that appear in the course of moving the colored squares from one spot to another by rotating the tiers of the cube in any specific series of sequences. So how do we develop a strategy that will increase the odds of solving the cube in a reasonable amount of time? Fortunately, by using the principles of symbolic representation, symmetry, relativity, spatial visualization and a little intuition, we don't have to try every one of the forty-three quintillion, two hundred fifty-two quadrillion, three trillion, two hundred and seventy-four billion, four hundred and eighty-nine million, eight hundred and fifty-six thousand (43,252,003,274,489,856,000) possible configurations that we might encounter on the path to solving the cube, and that's a good thing, but it should be obvious that the *optimum path* to a solved cube from any given scrambled state may still be very hard to find. But, don't despair, there are many less-than-optimum, but reasonably do-able ways to solve a scrambled cube.

Now that we have seen that the cube is actually composed of four different kinds of oddly shaped pieces, all connected in the center of the cube; not 27 mini cubes, as we may have thought, and have calculated the actual number of possible configurations, any one of which we may have to start with on our quest for a solution, we need to develop an unambiguous notation that is as simple as we can make it, and yet sufficiently complex for use in describing all of the possible rotations and states of a cube with nine rotational planes.

Notations and Algorithms

The notation generally accepted by most cube enthusiasts is described as follows: Facing any given side of the cube, the front, or side facing you, is indicated by the label **F**, the left side is labeled **L**, right side: **R**, upper side: **U**, back side: **B**, and bottom side: **D**. The bottom side is labeled **D** for "down" to avoid confusion with **B** for "back". Further, for algorithms describing rotations in the nine planes simulated by the cube: **F, L, R, U, D,** and **B**, are used to denote clockwise rotations, and **F', L', R', U', D',** and **B'**, denote counter-clockwise rotations. (*Clockwise* and *counter-clockwise* are defined as if you were facing the side being rotated.)

You may be happy to learn that **D, D', B,** and **B'** are not needed in the set of instructions developed in this book, that are designed to solve the cube. But they must be included in any complete description of moves and may be needed in complex simulations using the cube. In addition to **F, L, R, U, F', L', R',** and **U'**, only one other symbol is needed, and that is the number 2 placed after a letter to indicate 2 rotations. There is a simpler, more efficient notation using only the cardinal numbers 1, 2, 3, 4, 5, 6, 7, 8, 9, and zero. But I am using the alphanumeric notation described above because I believe the acronyms help us to visualize and remember the algorithms more easily than the numeric notation.

In mathematics, a series of instructions for performing a specific mathematical operation is called an *algorithm*. Similarly, for the cube, an algorithm is a series of ninety-degree (quarter-turn) rotations that will change the cube in some specific way. For example, **LU'R'UL'U'RU2** instructs you to perform seven specific ninety-degree rotations and one 180-degree rotation, in order reading from left to right. Under appropriate circumstances, in the process of "solving" the cube, performing this algorithm will align the corners in the upper face of the cube.

The point of all this is that without effective algorithms, solving the cube might prove to be practically impossible. So, the question to ask is: is it possible to develop an algorithm to solve *any* scrambled cube, and if so, how do we go about finding it? The answer for how to find it turned out to be very simple and basic: Look for patterns that resonate, by that I mean patterns that re-occur on a regular basis in the process of rotating the nine planes of the cube.

An important point to notice here is that this is exactly what we do in every field of science and in fact, in everyday life. We look for recurring patterns to gain an idea regarding what might happen next.

In science we look for recurring natural patterns, develop hypotheses that appear to explain them, express our theories in the objective language of mathematics, and then test them to see if they work repeatedly and consistently. If they do, they become known as "Laws of Nature". In everyday life, we try to behave in ways that will accomplish desirable goals, and we may want to change our approach when what we find that what we are doing fails to work and/or brings us displeasure or pain. If we make decisions that do not advance us toward our goal, we can change what we have been doing. As you may know, continuing to do the same thing over and over expecting different results is said to be 'one definition of insanity'.

Recurring patterns in the configurations of the cube, as in nature in general, result from the innate geometric structures of reality in the forms of matter, energy, time and space, and the effects of our actions as conscious beings. This is why the

cube can be used to model reality at every level, including the quantum, macro and cosmological levels.

Patterns

When I tackled my first cube, nearly 40 years ago, like most beginners, I found that I could easily maneuver the moveable pieces to make one side of the cube all one color by trial and error, and then move the edge pieces to "solve" one layer of the cube. I also found that this could be done more quickly and efficiently if I focused first on forming one specific pattern: a monochromatic (single colored) "X" on one side of the cube. I chose to form this X with the cardinal color red. See **Figure One**.

By not worrying about where any of the other pieces were while forming the X by moving the four two-faced *edge* pieces into place around the single-faced *center* piece, I could avoid counter-productive moves. Once the X was complete, the three-faced corner pieces were easily moved into place with just two or three rotations each, by paying attention to the colors and lining the edge pieces up with the center pieces of the same color in the second layer.

Moving on to solve the second layer without disturbing the first one appeared, at first, to be impossible. However, the lesson of focusing on patterns, learned in the process of completing the first layer, applied to moves involving the second layer, allowed me to overcome this barrier. Within a week, I was able to complete the first two layers, regardless of the scrambled configuration with which I started.

It occurred to me that because the center piece in each face never changes location relative to the other center pieces, that when I lined the ends of the X of the completed first layer up with their corresponding center pieces of the same colors in the second layer, I already had four of the eight pieces of the second layer in place. In other words, the second layer was already half solved just by rotating the X until it lined up with the center pieces! See **Figure Two**. The next step was to replace the two-faced edge pieces of the wrong colors in the second layer with the two-faced pieces of the right colors.

Because of the fixed geometry of the cube, all I had to do was find a series of moves, i.e. *an algorithm,* that would move a misplaced two-faced edge piece from the upper layer into its correct position as an edge piece in the second layer with proper color orientation, while returning any first-layer pieces that might be disturbed in the process back to their solved positions. This was easier than I expected because the overall parity of the cube is even. (I will discuss parity, combinations and permutations in more detail as we proceed.)

Because of parity, relocating one piece automatically relocates a second piece in a complementary way. By paying close attention to the orientation of the colors involved and the patterns of their spatial alignment, I found the algorithm that completed the second layer within a few hours. It has a right-handed version and a left-handed version, depending on whether the piece to be moved is on the right-hand or left-hand side of the edge you need to replace. Here are the moves:

Right: U'L'ULF'LFL' and **Left: URU'R'FR'F'R**.

This may be a bit confusing at first, but with practice, a somatic (muscle) memory is established and switching from left to right as appropriate becomes as easy as riding a bicycle.

PART I: KNOWING YOUR CUBE ～ 13

FIGURE ONE

14 SECRETS OF THE SACRED CUBE

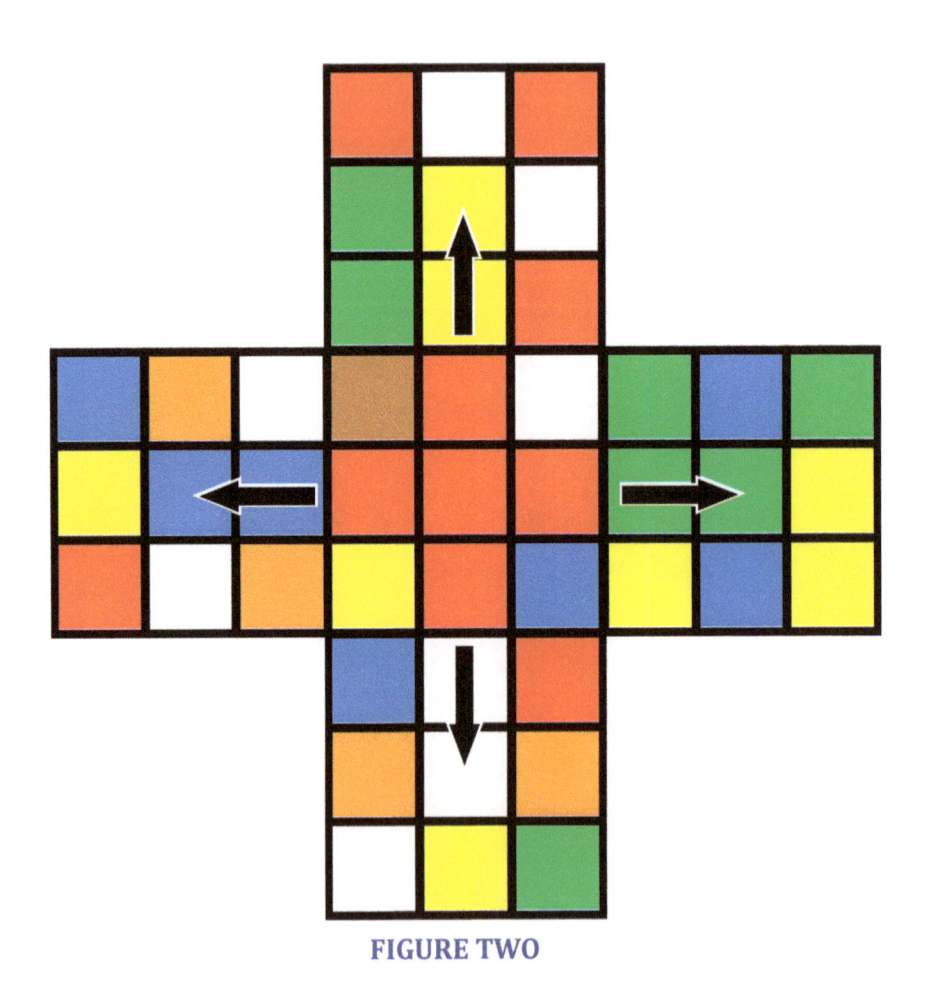

FIGURE TWO

Holding the cube with the completed first layer as the bottom layer, find a second layer edge piece that needs to be replaced because the colors are completely wrong (not just reversed). Then find the edge piece with the colors needed to replace it in the upper layer. Turn the upper layer of the cube until the front colored face of the edge piece to be moved into place is aligned with the front center piece of the same color. See **Figure Three**. If the edge piece to be moved into place is on the right of the corner, use the right–handed version above, if it is on the left of the corner, use the left-handed version. This will move the replacement piece into place and return the first layer (the bottom layer) to its solved state with the second layer edge replaced by the correct piece. If a second layer edge has the correct piece in place where it should be, but with its two colors reversed, use this algorithm twice: first to remove and replace the piece with any upper layer edge piece, and then, after realigning it, to replace it with the correct edge piece in the proper orientation.

As you might guess, solving the third layer without scrambling the first and second layers, is more difficult. Recognizing patterns and knowing what to do when they appear, is again the key. With the first and second layers completed, one of four patterns will be found on the face of the third layer. The pattern will be in the color opposite of the color chosen for the first layer. On my cube, orange is opposite the color red, which I chose for the face of the first layer. You need not be concerned about the color and location of any other pieces in the upper face at this point.

The four patterns possible are shown in **Figure Four**. In the very rare case that you have pattern A, _and_ all the corners are in place with the orange color up, the cube is solved. In cases of pattern A where the corners are not in place, use algorithm A to align the corners. If B is the pattern you see, use algorithm B, if the pattern is C, perform algorithm C. Finally, if the pattern is D, use algorithm B or C to obtain pattern A, B or C and then proceed as described above for those patterns,

Algorithm A: **LU'R'UL'U'RU2**

Algorithm B: **FURU'R'F'**

Algorithm C: **FRUR'U'F'**

PART I: KNOWING YOUR CUBE 17

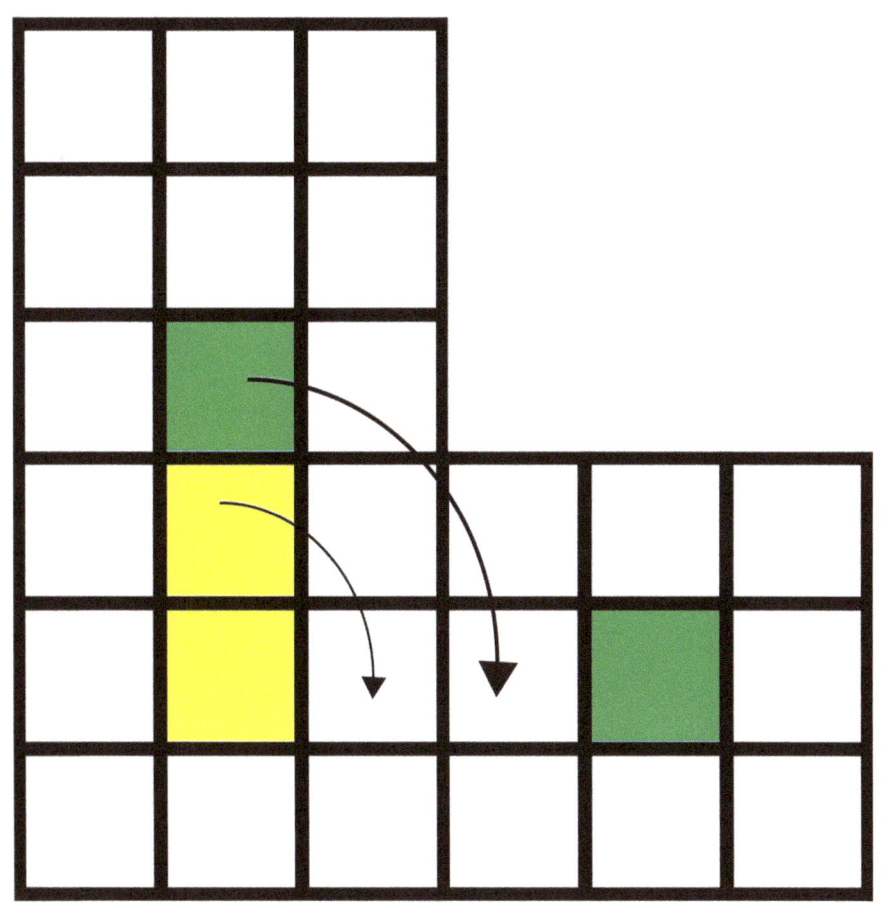

FIGURE THREE

18 ∞ SECRETS OF THE SACRED CUBE

A. LU'R'UL'U'RUS

B. FURU'R'F'

C. FRUR'U'F'

D. Apply B or C

FIGURE FOUR

You should now have the corners of the third layer in place, with layers one and two complete. If not, you have slipped up somewhere and will have to recover by backing up or starting over. Warning: even the slightest slip up, like skipping a rotation, or performing a rotation in the wrong direction, will abort the solution and the cube will, in effect, be re-scrambled. But if you have made your way carefully to the point where you have the corners in place, and the first two layers are still intact, you will see one of the eight patterns shown in **Figure Five**. The arrows indicate *orange* faces and the directions they are facing. Keep in mind that the patterns shown in **Figure Five** are on the upper face of the cube, U, and be sure to orient the cube with the front, F, the bottom in **Figure Five**, facing you, to execute the rotations of the next set of algorithms.

If you have patterns 1 or 2, moving to completing the solution *can* be relatively easy. If you have pattern number 1, use algorithm D; if 2, use algorithm E.

Algorithm D: **R'U'RU'R'UR2U2**

Algorithm E: **RUR'URUR'2U2**

After applying algorithm D or E, to pattern 1 or 2, there are 3 possibilities: 1,) The cube is solved; 2.) The cube is not solved but can be solved by repeating D or E one or more times or; 3.) The edge pieces will need to be rotated around the upper face to complete the solution, in which case, use algorithm F to rotate them clockwise, or algorithm G to rotate them counterclockwise, as needed. Case 1 and 3 lead directly to solution, but case 2 may lead to another of the patterns, where application of the algorithm for that pattern will be necessary.

Algorithm F: **R'U'RU'R'UR2U2** (to rotate clockwise)

Algorithm G: **RUR'URUR'2U2** (to rotate counter-clockwise)

Before I go into the details of how to deal with patterns 3 through 8, I want to bring to your attention what I consider to be three of the most important things to be learned from solving the cube.

These three things have to do with *focus*, *memory* and *intuition*. Laser-like *focus* is needed, especially when you are just starting to learn to solve the cube, before the patterns are fixed in your mind. Whether you realize it or not, you are equipped with several kinds of *memory*. In solving the cube, you must consciously and deliberately use at least two of them: mental memory, and somatic, or muscle memory. You must first use mental memory to memorize the algorithmic acronyms needed to solve the cube, and then practice performing them until somatic memory is established to the point where your hands know how to perform them without you having to think about them. This is similar to what happens when you learn to ride a bicycle, swim, play a musical instrument, learn a language, or learn to perform any action where you either cannot, or do not want to take the time to translate a symbolic mental memory into physical action.

PART I: KNOWING YOUR CUBE 21

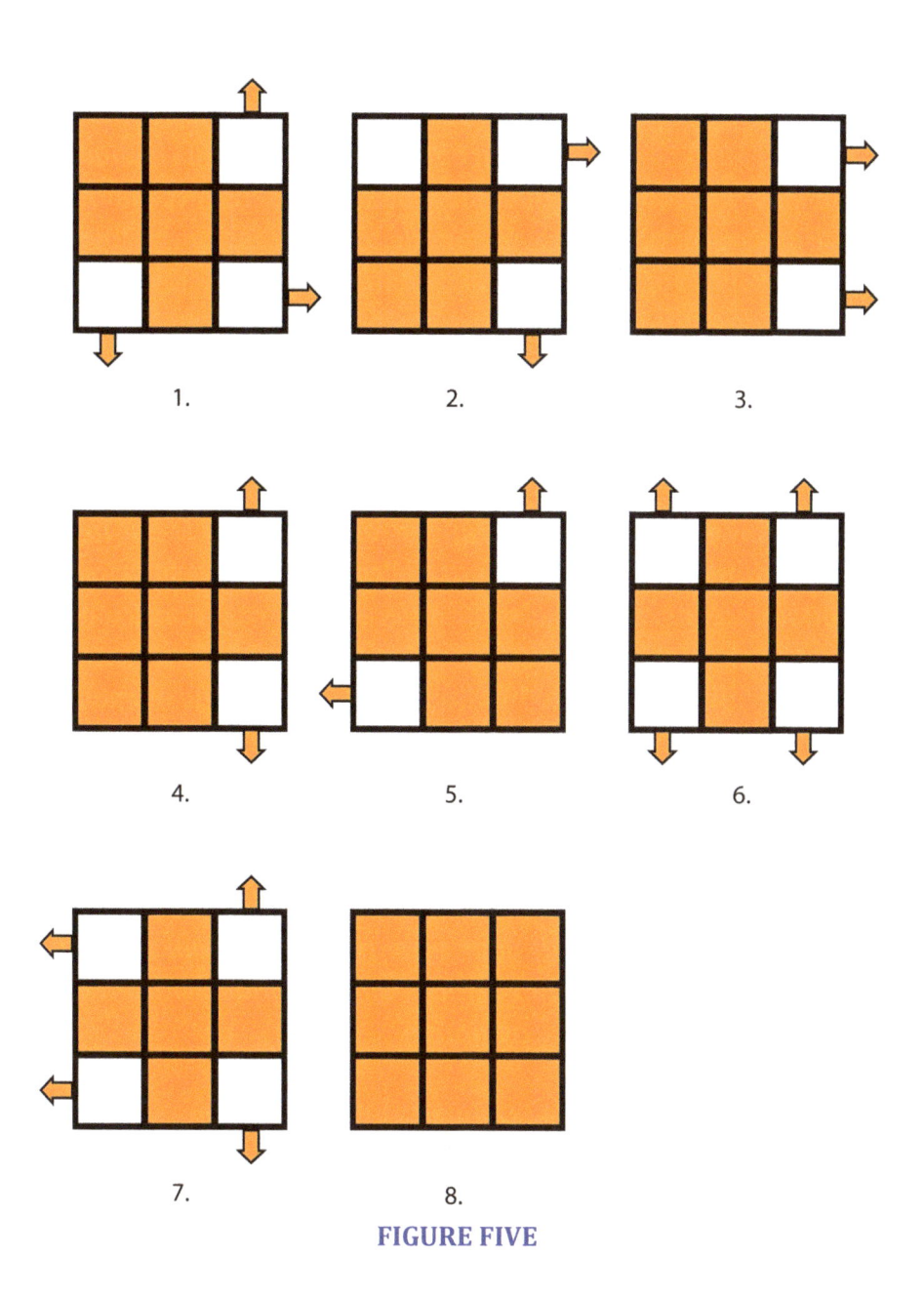

1.   2.   3.

4.   5.   6.

7.   8.

FIGURE FIVE

Interestingly, somatic memories of physical actions performed may remain long after the conscious memory of symbols representing them have faded. After solving the cube in 1981, I didn't pick it up again for several years. When I did, the notes I made in 1981 had been lost in one of our moves, and I couldn't remember the algorithms. But when I relaxed, just started twirling the cube and let my muscle memory take over, I was able to solve it again, even though at first, I couldn't slow down enough to write the algorithms down without losing track of the thread of logic linking the moves.

Intuition is an even more subtle form of memory that involves the recognition of innate patterns in reality that repeat themselves under certain circumstances. **Real intuition is not just lucky guesswork, it operates at the conscious, subconscious and superconscious levels**. The difference between conscious, subconscious and superconscious intuition and how to access them is a subject for another book, but, like somatic memory, intuition is available to anyone who has the patience to work hard enough to learn how to use it. That ability is suppressed by most people because we are almost always focused on localized mental or somatic activity.

If you can learn to shift your attention away from mental and physical activity and refocus on conscious awareness, intuition becomes more available.

There are reports of very young children solving the cube in less than a minute. In some cases, these are tricks accomplished by starting with a specific configuration for which a solution has been worked out and memorized. But it *is* possible that some young children may be able to solve a random scramble, because young children, unlike most adults, have not yet cluttered their minds with meaningless mental chatter and emotional attachment to temporal memories and desires for future outcomes. And it may be that such savant children are aware of universal patterns that get obscured by mental and emotional clutter as they age, and are all but completely obliterated by strong physical and emotional attachments when they enter puberty.

Back to the cube: because the uncolored faces in the patterns of **Figure Five** may be any of the five remaining colors, there are a number of possible color combinations for each of the eight patterns. The intuitive perception of the logical sequence leading to solution of the cube in each case is a powerful alternative to attempting to solve the cube from that point by trial and error, checking out each sequence until you find the one that solves the cube. So, to solve the cube from these patterns, you can either develop algorithms for every possible permutation of every possible pattern or learn to use your intuition. If we choose to continue developing algorithms as we have been doing so far, then, since the cube is a finite physical object, there are a *finite, but large, number* of possible combinations at this point for which algorithms can be identified and memorized.

For all eight of the patterns in **Figure Five**, we have the lower layers of the cube solved and we know where all nine orange faces are in the upper layer. They are indicated by the arrows in **Figure Five**. But each of the faces left blank in the patterns may be any one of five colors. To find out how many different configurations are possible with the blank faces filled in in each of the eight patterns, we must return to the world of *permutations and combinations*. The formula for the total number of possible permutations of n things is n! So, for example, since 5! = 1x2x3x4x5=120, five colors can be arranged 120 different ways. If we want to know the number of possible permutations of n things taken k at a time, the formula is:

$$P(n,k) = n!/(n-k)!$$

The formula for the number of ways **n** things can be combined **k** at a time, called combinations, is **C(n,k) = P(n,k)/k!** or, by substitution of the formula for **P** in this equation: **C(n,k) = n!/(n-k)!k!**

And thus, the number of different configurations possible for pattern #1 is the number of different combinations of five colors taken three at a time:

C(5,3) = 5!/(5-3)!3! = 120/2x6 = 10 different configurations

Applying the formula for combinations to the eight patterns, we see that there is a total of 60 different configurations possible. With the algorithms we've used to get us to this point, and if we have to develop an algorithm for each of the eight patterns with final rotational adjustment algorithms like D and E, we could have 70 algorithms or more. That's a lot of algorithms to memorize. We can shorten the list a lot by recognizing repeating patterns and using intuition.

If, rather than arriving at pattern number 1 or 2, you have one of the patterns 3 through 7, the path to completion is a little more complicated. This is where you can begin to develop or sharpen your abilities of memory, pattern recognition and intuition. When confronted with one of the patterns 3 through 7, relax and ask yourself which seems right, or more comfortable to you, algorithm D or E. Then execute your choice except for the final U2 and look for patterns. If this takes you in a loop

PART I: KNOWING YOUR CUBE 25

back to the pattern you started with, reverse your choice of D or E. Continue this procedure until you obtain either pattern 1 or 2, and then solve using the algorithms given for them above.

Finally, if you are confronted with pattern number 8, where all of the pieces in the upper layer have their orange faces up but have either the edge or corner pieces in the wrong locations, use algorithm A to align corners, and then use either F or G to rotate edges to their proper positions. If the problem is corners, one application of algorithm A may resolve the problem and produce a completed cube, or it may result in taking you in a loop back to one of the other patterns. On the other hand, if the problem is the edges, one or two applications of F or G, depending on which direction the edges need to be rotated, clockwise or counter-clockwise, will suffice to complete solution of the cube.

While applying this approach to solving the cube, you may begin to see shortcuts, and novel applications of the algorithms that are more efficient than those provided here. An important part of the learning process while solving the cube, is discovering that these algorithms have multiple applications, and developing the ability of knowing when and how to use them is even more important. With practice, your intuition and ability to recognize patterns in advance may improve to the point where you will be able to avoid time-consuming loops and be able to solve the cube from any random configuration in less than one minute.

Tricks and Illusions

I can't move on without mentioning that there are many other ways to solve the cube, including those on the official Rubik's cube website. And I must warn you about a couple of approaches that you should avoid. They are tricks that are not really solutions. One trick that can enable you to impress your friends is the following: Prepare a new or previously completed cube in advance by scrambling it any way you like and recording the rotations you make. Then memorize them in reverse, and you have an algorithm that will solve that scrambled cube. A cube can look thoroughly scrambled with as few as 15 or 20 quarter rotations. When you have the moves thoroughly memorized, show your friends the scrambled cube, and proceed to "solve" it very quickly by performing the rotations that scrambled it in reverse. Of course, this is not really solving the cube. It's just a trick.

There's another trick that you <u>*really*</u> want to avoid, because it may ruin your cube: Some people, frustrated by months, or maybe even years of trying to solve the cube, have resorted to peeling the colored squares off the faces of their scrambled cube (which you can do with some of the Rubik's cube copies that are made that way) and gluing them back on to produce a "solved" cube. Wanting to do this is understandable, but it is very likely to result in making a cube treated this way completely unsolvable the next time it is scrambled.

Starting with a random scramble, the chance of the peeled and replaced cube being solvable when scrambled again, is no better than one in twelve. How can that be? It has to do with the geometry of the cube and the basic mathematical concepts of *permutation* and *parity*. For example, with a cube that has been altered to "solve' it, a person who knows how to solve the cube may find that, when he gets to the final stage of solving the cube, only one sub-cube is in the wrong location. Such a cube is unsolvable because of the geometrical parity of a 3x3x3 cube. Any algorithm that moves a sub- cube from one location to another, replaces another sub-cube when it moves the cube that is out of place. Thus, moving the cube to its rightful location dislodges another cube and moves it to an unsolved position, because of the parity of the cube as a connected system of moveable objects.

As I said, there other methods for solving the cube, but most legitimate methods will include some or all of the algorithms presented here in some form. You may want to explore some of the other methods if they appeal to you, but it is best to pick one method and stick with it until you have mastered it.

Combinations, Permutations and Parity

We used the formulas for permutations and combinations above without fully explaining why two different formulas are needed, and how they differ. Unexplained mathematical formulas always make things sound complicated, but this is really not that complex. *Permutation* is just a word that mathematicians use to distinguish the arrangement or configurations of things where the order of the arrangement is important, from situations where order is not important, where the word combination is used. And *parity* is just a word that mathematicians use to describe whether the count of a number of possible rotations is odd or even.

Parity is a very easy concept to understand, while understanding the difference between permutations and combinations is a little more difficult for most of us. Much of the confusion is caused by the fact that in common usage we generally call both ordered and non-ordered arrangements *combinations*. In fact, they are mathematically different, but not independent of each other. Permutations are *ordered* combinations.

It will help to look at permutations and combinations in specific concrete examples. If we count every possible grouping and ordering of a given number of things as a permutation of those things, as we do when arranging a specific number of different colors, numbers, or anything else, in every way possible, we have 1x2x3x4x...xn permutations, expressed symbolically as n! where n is the number of things we are permuting.

To see why this is so, start with 1 thing, then 2, then 3, and so forth, arranging them at each step in every way possible. You will find that the number of permutations for 2 things is 1x2 = 2, the number of permutations with 3 things is 1x2x3 = 6, etc. As you increase the number of things being combined you will see that the number of possible arrangements counted for n things is equal to the number of permutations for n-1 things multiplied by n, which is generalized in the formula P = n!

For example, if we have ten objects, like the ten digits: 0, 1, 2, 3, 4, 5, 6, 7, 8, 9, arranged in every way possible, we have 10! = 1x2x3x4x5x6x7x8x9x10 = 3,628,800 different permutations; but if we want to combine ten things in only one specific way, like in groups of 7 digits where the order of the 7 digits is important, as in conventional phone numbers, or the arrangement of 9 small squares of each of six different colors on the surface of a cube of 54 squares, we see that, as each distinct group is chosen, the total number of possible additional groupings is reduced.

For phone numbers the possible permutations for the 10 numbers 0, 1, 2, 3, 4, 5, 6, 7, 8, 9, in groups of 7, is P = n!/(n − k)! where n is the number of things the groups are being chosen from, and k is the number of things in each group selected. So, the number of unique phone numbers possible in one area code is P =10!/(10-7)! = 604,800, while the number of possible combinations of 10 digits taken seven at a time, without regard for order, is C = P/k! = n!/(n – k)!k! = 3,628,800/(3!x7!) = 3,628,800/(6x5040) = 3,628,800/30,240 = 120.

If, even with these examples, permutations and combinations are still a bit confusing, an overview may be helpful: In these examples we have described three different ways of arranging 10 things. In the first, we are counting all of the possible arrangements or ordered lists of numbers as different combinations called permutations. Order is very important in permutations. For instance, there are 3,628,800 different ways or sequences in which the 10 ingredients of a chocolate cake (flour, sugar, milk, eggs, salt, shortening, chocolate, baking powder, vanilla, and heat) can be combined, but only one of them is likely to produce the cake you are expecting. Some of the results of other sequences may be edible, and some will not be very appetizing.

In the second way, we are counting lists of a specific number of things that can be combined from a list of a larger number of things. Order is still important, so the members of this list are also called permutations, but order is not as important as in the first way of ordering 10 things, so the number of permutations will be reduced. An example would be creating a list of all of the seven-digit phone numbers that can be formed from the nine cardinal numbers and zero, as we did above.

In the third way of combining 10 things, there is no concern about order, the combinations are not ordered combinations, so they are not called permutations; they are just called combinations. An example might be combining 10 kinds of fruit to make a fruit salad. They can be combined

120 different ways, but the result will always be the same fruit salad.

Now, with these concepts of combinations, permutations and parity in mind, let's look at your cube. There are eight corner pieces with three colored faces, twelve edge pieces with two colored faces, six center pieces with one face each, and every face has one of six colors. The parity of the colors of each corner piece is odd (3), and the parity of the color of each center piece is odd (1), while the parity of each edge piece is even (2). Since there are an even number of pieces with odd parity, and no face can change its color, the overall parity of the cube is always even, even though the parity of parts of it by themselves may be odd. Knowledge of this fact is useful in choosing which algorithm to use in a given pattern.

Notice that every algorithm is made up of clockwise and counter-clockwise rotations, and every rotation moves pieces from one location to another location, while the center pieces are never moved from their locations, they just rotate in place. So, the parity of the faces moved in any specific rotation or series of rotations is always even. On the other hand, the parity of the upper layer of the cube, with 21 faces is odd.

Now look at the cube when you have the first two layers completed but have the third layer yet to be solved. Any algorithm that advances the cube toward completion by relocating some of the faces in the third layer, will have to move some of the pieces in the first two layers in the process, but it will also have to return them to their solved positions when the algorithm is completed. This means that we can determine what can and can't be said with regard to parities in the third layer before and after application of that algorithm without regard to the movement of pieces in the first two layers.

So, considering the third layer by itself, we see that the face and color parities are all odd: There are 21 faces, and 5 colors. Yes, 5, not 6, because the nine faces of one of the six colors, that's red in the case of the way I position my cube, will all be in place on the bottom of the cube, and there will be exactly 3 blue, 3 green, 3 yellow, 3 white, and 9 orange faces, in some configuration in the third layer. Because of the geometry of the cube, the numbers of faces of each color in the final layer is invariant; that is to say, they are the same, regardless of how we arrived there.

Now it is easy to see that if we find certain configurations of colors in the third layer after the first two layers have been solved, it indicates that the cube has been tampered with, and can't be solved. For example, if only one piece is out of place, because of parity, no move or series of moves will solve the cube, because moving it to the correct location in the third layer will displace one of the other pieces that was already in the correct position to an incorrect position.

Also, the permutations of an odd number of pieces is odd, requiring an even number of moves in the algorithm to correct them, which would violate the parities of the layer, and therefore can't be produced by any number of rotations from the solved state. This means there are at least $1 + 3 + 5 + 7 + 9 = 25$ configurations that cannot occur in the third layer unless the cube has been altered in a way that renders it unsolvable.

Gödel's incompleteness theorems tell us that there can be unsolvable questions in any consistent finite logical system, So, finite man-made structures like the Rubik's cube can be unsolvable, i.e., we may be unable to align such structures with any sort of logical reality. On the other hand, we know intuitively that the problems of an infinite reality are ultimately solvable because the system can always be expanded to include a new set of axioms. Otherwise, science would be pointless, and life would be frustratingly random and meaningless.

One of the most important discoveries of science was Max Planck's discovery more than 100 years ago that everything in the physical universe is meted out in multiples of a basic quantum unit. This means that there has to be a basic building block, an indivisible unit, combinations of which make up everything we can observe, weigh and measure. Thus, there must be an ultimate smallest quantum unit. The observable part of the Rubik's cube is made up of basic units of three types. Could reality be made of three types of quantum units? Turns out, it is!

We've solved the mysteries of the cube. We know how to take a completely scrambled cube in any one of the more than 42 quintillion mixed-up configurations and make it chromatically coherent in a reasonable length of time. Can we do the same thing with the seemingly random information of the physical world? Can we solve the puzzle of the physical universe? Can we understand reality in a way that makes everything make sense? Can understanding the cube be helpful? The answer to all of these questions is yes! Please read on to learn more.

PART II: SOLVING THE MYSTERIES OF QUANTUM REALITY

Reality is Quantized

Let's see how what we've learned about the cube can be used to help us understand the nature of quantum phenomena. The fact that we exist in a quantized reality, proved true for mass and energy by Max Planck more than 100 years ago, and expanded to include space, time and consciousness by several scientists, including Neppe and Close in 2011, provides the basis for a comprehensive model of reality with all finite phenomena describable as combinations of standardized quantum equivalence units, the true quantum units capable of describing the building blocks of physical reality.

For the moment, let's visualize the building blocks of reality as cubes consisting of combinations of the "small cubes" which were conceptualized as the 27 idealized cubes of the Rubik's cube. The individual small cubes represent quantum equivalence units. Certain combinations of these cubic units are symmetric about all three axes of the six possible mutually orthogonal directions from their centers. This means that such combinations will be stable as spinning objects, like the elementary entities of particle physics. Then we can represent an almost infinite number of combinations of spinning cubes, the most stable of which will also be cubes.

Next, suppose the spinning cubes are held together by a balance of centrifugal and centripetal forces; now we have an analogy to the Rubik's cube, which is held together by the triaxle at its heart. Because they are spinning, elementary particles have electric and magnetic polarity. To maintain stability, these polarities have to be orthogonal; otherwise, the particle would become asymmetric, causing it to wobble and fly apart. We can let the red-orange axis represent the electric polarity, and the blue-green axis represent the magnetic polarity of the spinning cubes.

The choice of colors is of course arbitrary, except that the orthogonal axial parities must be maintained to reflect the fact that electric and magnetic forces are orthogonal (acting at right angles to each other) in physical reality. This simply reflects the parsimonious symmetries of nature. As I proceeded in this manner, I discovered that the third polarity (the white-yellow axis) also represents an important feature of reality. These polarities represent three sets of parities: electric, magnetic and gravitational.

Why elementary particles are spinning at extremely high rates of angular velocity is also explained by the TDVP model, but that is a subject a bit beyond the scope of this book. We know, however, from empirical data, that elementary particles are indeed spinning. Are they spinning cubes? Probably not, but for maximum stability, they must be symmetric. However, as we shall see as we proceed, the shapes of the elementary particles that combine to provide the structures of sub-atomic reality make no difference, as long as the shape is the same for all of the particles being combined, and symmetric about the axes of rotation. Experimental data verifies this similarity of shape in elementary particles, and some well-established basic number theory and geometric theorems guarantee this mathematically.

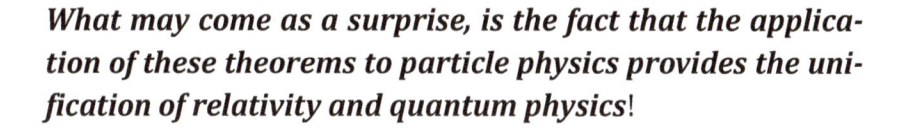

PART II: SOLVING THE MYSTERIES OF QUANTUM REALITY   35

What may come as a surprise, is the fact that the application of these theorems to particle physics provides the unification of relativity and quantum physics!

Simulation of the Combination of Elementary Particles

After two years of communicating by email, telephone and Skype, Dr. Neppe and I finally met in person in Amsterdam in 2010. We were discussing our theoretical model, the Triadic Dimensional, Vortical Paradigm when Dr. Neppe expressed the thought that the fact that quarks only combine in threes to form protons and neutrons might be another reflection of the basic triadic nature of reality. I agreed and added that I could explain why quarks combine in threes. I had intuited this fact and formulated a mathematical demonstration of the necessity of quantum distinctions combining in triads while applying the calculus of distinctions to the Pythagorean Theorem and Fermat's Last Theorem in 1965.

I have published proofs of the impossibility of merging two quantum distinctions to form a third quantum distinction, and the possibility and the proclivity of three quantum distinctions merging to form a fourth quqntum distinction in a number of places. including the book "Reality Begins with Consciousness", coauthored with Dr. Neppe in 2011, several journal articles and on my Transcendental Physics website www.ERCloseTPhysics.com. It is also a key element in my chapter of the book *Is Consciousness Primary, Perspectives from the Academy for the Advancement of Postmaterialist Sciences*, in press for publication in 2018 or 2019. The following link will take you to a post in my transcendental Physics site:

https://www.erclosetphysics.com/2015/03/science-and-spirituality-tv-new-youtube.html?showComment=1546228549960&fbclid=IwAR2i96xqyjmo7FRSiBgv9sZpNmgMOB1PcEoBUUUd1dO-DTHWv35jm-pL4xE#c2517893842173152298

or use the URL address https://youtu.be/lGOrg15HcRQ to go directly to some You-tube presentations Jacqui and I did supporting the process of putting Consciousness into the equations of physical science.

How do elementary particles combine to form physical reality? When drawn together by the forces of their electric and magnetic polarities, do elementary particles simply stick together as suggested by the tinker-toy diagrams in our physics and chemistry textbooks, or do they actually merge *volumetrically* like small drops of water merging to form a larger drop? It is obvious that the merging of spinning objects would produce a much more stable spinning object than objects simply stuck together, forming lumpy objects. In order for quarks to form protons, the most stable objects in the universe, they must merge volumetrically.

Unless the spins cancel because they happen to be equal and opposite, a particle formed by combining spinning particles will also be spinning, because the principle of conservation of mass and energy guarantees that the spin of the elementary particles, measurable as angular momentum, is passed on to the new particle created by the combining or merging of particles. Since we are exploring the ways we might be able to apply what we've learned with the cube, let's consider for the moment, that elementary particles might be spinning cubes.

PART II: SOLVING THE MYSTERIES OF QUANTUM REALITY   37

The space occupied by a spinning cube is spherical, somewhat larger than the volume of the cube. But the circumference, and consequently the volume of a cube spinning at a significant fraction of the speed of light relative to the observer, is reduced by the Lorentz contraction factor, which is the mathematical representation of the well-documented relativistic shrinking of the measurement of space due to motion relative to an observer.

When an electron in orbit is stripped from the outer shell of an atom, the volume it occupies decreases greatly, and the conservation of angular momentum causes its spin velocity to increase rapidly, like when a spinning skater pulls her arms in. Calculations prove that the spin velocity of the electron will reach light speed before the volume of the electron shrinks to zero. Thus, as the accelerating spin of the electron increases, its volume reaches a minimum value. We can take this volume as the volume of the quantum equivalence unit, and the mass of the electron, the smallest mass of the elementary particles making up normal hadronic matter, as the quantum equivalence unit of mass, and the MeV/c² energy equivalence of the electron mass as the quantum equivalence unit of energy.

The expression

$$\sum_{i=1}^{n}(X_n)^m = Z^m$$

with all variables and indices restricted to integers, represents a family of Diophantine equations describing the combination of **n** objects: $X_1 + X_2 + X_3 + ... + X_i + ... + X_n$ of dimensionality **m** to form a composite object, Z^m. (A Diophantine equation is simply an equation for which integer solutions are required.) When the integers X_i are the numbers of quantum equivalence units representing the mass/energy of elementary particles, we have an expression that can represent any combination of particles. Because the various forms of this expression as **m** varies from 3 to 9 conveys the geometry of 9-dimensional conscious substrate of reality to our observational domain of 3S-1t, we call this expression the "Conveyance Expression", and individual equations produced by the expression "Conveyance Equations". When **n = m = 2**, the expression yields the equation

$$(X_1)^2 + (X_2)^2 = Z^2$$

which, when related to areas, describes the addition of two areas, A_1 and A_2 with sides equal to X_1 and X_2 respectively, to form a third area, A_3, with sides equal to Z. When these squares are arranged in a plane to form a right triangle, as shown below, we have a geometric representation of the familiar Pythagorean Theorem, demonstrating that the sum of the squares of the sides of any right triangle is equal to the square of the third side (the hypotenuse) of that triangle.

The Pythagorean Theorem

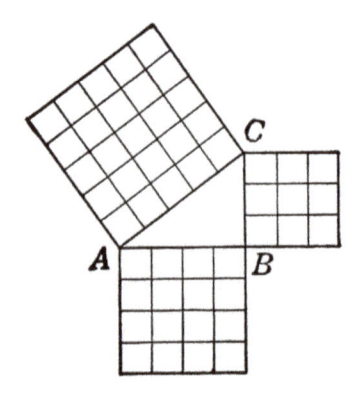

$$(AB)^2 + (BC)^2 = (AC)^2$$

PART II: SOLVING THE MYSTERIES OF QUANTUM REALITY　39

Considering the Pythagorean equation as a Diophantine equation, we find that there exists an infinite set of integer solutions with AB = X_1, BC = X_2 and AC = Z. Members of this set, e.g. (3,4,5), (5,12,13), (8,15,17), etc. yield $3^2 + 4^2 = 5^2$, $5^2 + 12^2 = 13^2$, $8^2 + 15^2 = 17^2$, ...

These triadic sets are called "Pythagorean triples". When **n = 2** and **m = 3**, the expression becomes the equation $(X_1)^3 + (X_2)^3 = Z^3$, where X_1, X_2 and Z represent the determinant measures of volumes, just as they represented areas when **n = m = 2**. Using the unitary quantum equivalence unit as the unit of measurement, we have a Diophantine equation representing the combination of two elementary particles. Their volumes are equal to $(X_1)^3$ and $(X_2)^3$, where X_1 and X_2 are whole number multiples of unitary cube volumes, analogous to the small cubes of the Rubik's cube.

The volumes *occupied* by these spinning objects are spherical, with radii equal to r_1, and r_2, and the volume of a sphere of radius r is $4/3π(r)^3$. So $(X_1)^3 = 4/3π(r_1)^3$ and $(X_2)^3 = 4/3π(r_2)^3$, and if they combine to form a third particle, represented by $Z^3 = 4/3π(r_3)^3$, we have:

$$4/3π(r_1)^3 + 4/3π(r_2)^3 = 4/3π(r_3)^3.$$

Dividing through by the common factor, $4/3π$, gives us:

$$(r_1)^3 + (r_2)^3 = (r_3)^3.$$

Since this equation is of the same form as Fermat's equation: $X^m + Y^m = Z^m$ when **m = 3**, **Fermat's Last Theorem** tells us that if r_1 and r_2 are integers, r_3 cannot be an integer. This means that if X_1 and X_2 are integer multiples of quantum equivalence units, then **Z** *cannot* be an integer multiple of the quantum unit. This means that the right-hand side of this equation, representing the combination of two quantum particles, cannot be a symmetric quantum particle. And, because Planck's discovery that our physical reality is quantized, no particle can contain fractions of mass and/or energy units, the volumetric merger of two symmetric particles will always produce an unstable asymmetric spinning object.

Notice that the *shape* factor **4/3π** cancelled out, indicating that this type of Diophantine (integer) equation is obtained regardless of the shape of the particles, as long as the shape is symmetrical and the same for all three entities.

A key element of TDVP is the understanding that when we change our focus from the macroscale to the quantum scale, we do not cross over into some different kind of reality governed by a strange new set of rules that are unknown and undefinable in the macroscale world. **There is only one reality; and there has to be one set of mathematically consistent laws governing it**. The quantum world only seems strange because we have failed to properly apply the basic relativistic and quantum principles in an appropriate way, to our observations and measurements.

What are the real differences between reality on the macro and quantum scales? One is, of course, the extreme difference in size. In our everyday interactions with realty we are not aware of the fact that everything is made up of quanta because the size of the quantum is so far below our ability to detect it that observations with our limited physical senses, and even, for the most part, through any existing physical extensions of them, cause physical reality to appear to be continuous.

Proof that the reality of the quantization of mass and energy discovered by Planck implies quantization of space and time as well, is beyond the scope of this little book, but there are no fractional quanta in the real world. To properly reflect this, we must change all of the equations describing reality to Diophantine equations. We can do this by replacing Newtonian calculus with a more basic calculus we call the calculus of dimensional distinctions. We will wait to discuss this in more detail later, because it is not as important in the present discussion as is the second difference between macro and quantum observations. The second, and most important difference, is that elementary particles are spinning very rapidly. When they reach angular velocities approaching the speed of light, volumetric space relative to them becomes quantized. At that point, spinning particles can combine, merging like drops of water. The volumes $(X_1)^3$ and $(X_2)^3$ combine to form a symmetrically stable spinning particle Z^3, only if Z is equal to a whole number of quantum equivalence units. If Z is not an integer, the resulting combination will be lop-sided and very unstable.

To help us understand why this is true, we can visualize our quantum unit as existing in the shape of the simple six-sided regular solid polyhedron, a cube. Using this quantum cube to simulate a quantum building block, we can construct symmetrical compound objects by constructing larger cubes with unitary cubes. A cube with two unitary building blocks on each side contains 8 blocks; a cube with three blocks on each side contains 27 blocks; a cube with four blocks on each side contains 64 blocks; etc. In general, a cube with N quantum blocks on each side contains N^3 blocks. But we must also realize that as observers at rest relative to these rapidly rotating objects, because of the relativistic contraction of the volumes of the spheres, our measurements of the diameters and radii of the volumes occupied by the spinning cubes are reduced accordingly.

Combinations of Elementary Particles and Fermat's Last Theorem

At the critical velocity, when the relativistic circumferences of the spinning objects have contracted to the point where they are equal to X_1 and X_2, the particles can merge, and $4/3\pi(r_1)^3 = (X_1)^3$, $4/3\pi(r_2)^3 = (X_2)^3$, and $4/3\pi(r_3)^3$ becomes Z^3 where, the resulting object, represented by Z^3, is symmetrical if, and only if, Z is an integer multiple of the quantum equivalence unit. But Fermat's Last Theorem tells us that Z cannot be an integer. Therefore, the combined high-velocity angular momentum of the asymmetric compound particle formed from two symmetrically spinning elementary particles will cause it to wobble, spiral and fly apart. This may lead us to wonder how it is that there are stable particles in the universe, and why there is any physical universe at all.

It is this requirement of symmetry for physical stability that *reveals* the intrinsic dimensionometric structure of reality that is reflected in the Conveyance Expressions. It turns out that there *can* be stable structures, because for **n = m = 3**, the Conveyance Expression yields the equation:

$$(X_1)^3 + (X_2)^3 + (X_3)^3 = Z^3,$$

which we find, **does** have integer solutions. The first one (with the smallest integer values) is:

$$3^3 + 4^3 + 5^3 = 6^3$$

We've seen that when **n = 2** and **m = 3**, **Fermat's Last Theorem** proves that two elementary particles cannot combine to form a stable compound object. On the other hand, the Conveyance

PART II: SOLVING THE MYSTERIES OF QUANTUM REALITY   43

Expression when **n = m = 3** has integer solutions, showing that trinomial combinations of elementary particles *will* form stable structures.

This explains why up-quarks and down-quarks only combine in triads to form stable protons and neutrons, without which there could be no stable atoms in the physical universe.

Most mathematicians and physicists regard theorems like Fermat's Last Theorem as pure number theory math with no practical application. This is an error in thinking perpetuated by academic specialization and reductionist approaches to the analysis of physical reality. Fermat's Last Theorem is a prime example of how a pure mathematical theorem can reflect the innate structure of reality.

By uniting the principles of relativity and quantum physics in the analysis of the electron and thinking of elementary particles as cubes, we have taken the first step necessary to simulate physical reality with Rubik's cubes.

The True Quantum Unit and The Third Form of Reality

The quantum unit is a sub-quark unitary extent/content unit of measurement for use in describing the combination of spinning particles. The cube, with nine tiers of nine unitary cubes, capable of rotation around a common center, provides a suitable analog model that is useful for simulating subatomic structures. This is especially true because the mathematics of the TDVP model requires nine dimensions to describe our experience of space, time and consciousness.

When we choose to measure the substance of a quantum distinction, the effects of its spinning in the three planes of space register as inertia or mass, spin in the time-like dimensional planes manifests as energy, and spinning in the additional planes of reality *containing* the space and time domains, requires a third form of the stuff of reality, in addition to, but not registering as either mass or energy, to complete the minimum quantum volume required for the stability of that distinct object. Because this third form of the stuff of reality is unknown in current science, we chose the third letter of the Aramaic alphabet, ג (gimmel), and we call the sub-quark equivalence unitary measure of the three forms of reality the ***Triadic Rotational Unit of Equivalence***, or **TRUE** Unit.

The mix of the three forms, mass, energy and gimmel (**m, E** and ג) *needed to maintain symmetric stability*, present in any given 3S-1t measurement, will be determined by the appropriate Conveyance Equation, as demonstrated below. When **n = m = 3**, $\sum_{i=1}^{n} (X_n)^m = Z^m$ yields:

$$(X_1)^3 + (X_2)^3 + (X_3)^3 = Z^3$$

The integer solutions of this Diophantine equation in TRUE units represent the possible combinations of three symmetric vortical distinctions forming a fourth three-dimensional symmetric vortical distinction.

Particle Physics and the Quantum World

The analysis of the double-slit and delayed-choice experiments, and the resolution of the EPR paradox, tell us that reality at the quantum level is like an all-encompassing interwoven multi-dimensional tapestry, but because of the extreme smallness of the quantized structure, far smaller than we are able to see directly, even with the best technological extensions of our physical senses, we are directly aware only of the broad-brush features that seem to exist as separate objects. We have tried repeatedly, over the history of modern science, to identify the most basic building blocks of physical reality, starting with large structures like cells, molecules and atoms, proceeding to smaller and smaller objects, only to have them slip through the finer and finer-scale net of our search. Relativity and quantum physics tell us, however, that there is an end to this, a limit to this infinite descent of spinning particles, a bottom to our search: the smallest possible particle, the minimum quantum equivalence unit.

The force causing spinning motions in the finite distinctions of physical reality is the continuous force of universal expansion. The fact that expansion is uniform and continuing, perhaps even accelerating, indicates that there is no physical structure outside the universe to impede or alter uniform expansion. As part of the nine-dimensional universe, the maximum expansion velocity between two farthermost separated points in a quantized 3S-1T reality is light speed, a speed determined by the mass/energy ratio in the observable universe: **$c = \sqrt{(E/m)}$**.

The mathematical expression of the conveyance of logical structure can be derived by application of the CoDD and Dimensional Extrapolation to the elementary distinctions of extent and content revealed by the empirical data obtained in particle colliders, under the integer requirement of quantization. Particle collider data provides us with an indirect glimpse of the origin of the elementary structures that make up the limited portion of reality observable in 3S-1t. Using particle collider data and the mathematical principles of quantum physics and relativity, we can derive the equations describing the combination of elementary particles to form stable sub-atomic structures. Because we exist in a quantized reality, these equations will be Diophantine equations, i.e. equations with integer solutions. We call the general mathematical expression summarizing these equations the *Conveyance Expression* because it contains within it the mathematical relationships that convey and limit the logical structure of the transfinite substrate through the sequentially embedded nine-dimensional domains of finite distinction to the 3S-1t domain of physical observation and measurement.

Within the framework of the current Standard Model of particle physics, the basic concepts of quantum physics and relativity are applied to the particle collider data to yield numerical values of the physical characteristics of the sub-atomic particles perceived to be the building blocks of the observable universe, including photons, electrons, neutrons and protons, in units of MeV/c^2. Analysis of these data in the framework of the mathematics and geometry of TDVP in 3S-1t provides us with a way to find the true quantum unit of measurement. The empirically measured and statistically determined inertial masses of the three most basic elementary entities believed to make up what we perceive in 3S-1t as matter, i.e. electrons, up-quarks and down-quarks, are approximately 0.51, 2.4 and 4.8 MeV/c^2, respectively. The values for up and down quarks are derived statistically from millions of terabytes of data obtained from high-energy particle collisions engineered in specially-built colliders.

It is obvious from these data that the conventional unit: MeV/c^2 is not the basic quantum unit, because the data expressed in these units contain fractions of MeV/c^2 units. Max Planck discovered that energy and matter occur only in integer multiples of a specific finite unit of quantum action, not fractions of units. Therefore, the masses of the electron, up-quark and down-quark should be integer multiples of the basic quantum unit of mass/energy equivalence. Since the masses are fractional in MeV/c² units, one MeV/c² must be a multiple of a yet smaller truly *quantum* unit.

Except for the electron, the data for the mass/energy of the up- and down-quarks, in Table One below, are presented as ranges of values because the mass/energy of elementary particles are indirectly determined as energy equivalents from particle collider detector and collector data. Some measurement error can occur in any experiment, and even with the advances in technological precision from the first "atom smasher", the Cockcroft-Walton particle accelerator, in 1932 to the Large Hadron Collider (LHC) today, some measurement error is still unavoidable due to the extreme smallness of the phenomena, the indirect and delicate methods of measurement and the statistical interpretation of the data. The electron mass is considered to be one of the most fundamental constants of physics, and because of its importance in physical chemistry and electronics, great effort has been spent to determine its inertial mass very accurately at 0.511 MeV/c².

The integer values in Table One are obtained by assuming that the electron has the least mass of any stable elementary particle and is the smallest sub-atomic particle. Setting its mass to unity and determining the average masses of the up- and down-quarks as multiples of that unit, we have the normalized masses of the electron, up- and down-quarks. Using the latest available collider data, the mass/energy averages for the up- and down- quarks are 2.01 MeV/c² and 4.79 MeV/c² respectively. Dividing by 0.511 and rounding the nearest integer value, we have the normalized mass/energy equivalence for the electron, up- and down- quarks, as 1, 4 and 9 respectively. Using these normalized values, we can investigate how as finite distinctions they can combine to form protons, neutrons and the progressively more complex physical structures that make up the Elements of the Periodic Table.

TABLE ONE:
Fermions The Most Common Particles comprising the physical universe

Particle	Symbol	Spin	Charge	Mass (In MeV/c^2)	Mass/Volume (Naturalized)
Electron	e	1/2	-1	0.511	1 *
Up quark	u	3/2	+2/3	1.87 – 2.15	4 **
Down Quark	d	3/2	–1/3	4.63 – 4.95	9 **
Proton	P+	1/2	+1	740 - 1140***	1035***

* The mass of the electron is Naturalized to unity (1).

**The masses of the quarks are normalized as integer multiples of the naturalized electron mass unit. This is justified because the actual values must be integer multiples of the basic unit of quantized mass.

*** The fact that the detected mass of the proton is nearly 100 times more than the combined mass of two up-quarks and one down-quark is explained in the Standard Model by gluons and bosons thought to exist in the space around the quarks, although they are not detectable until "teased" into existence by high-energy collisions in the LHC. This discrepancy is explained by spin dynamics in TDVP.

The smallest finite unit of volume is the smallest possible distinction of extent that can be occupied by an accelerated spinning vortical object. This distinction of extent has a finite value because of the limit placed on the rotational velocity of any object possessing inertial mass by the light-speed limit of relativity. As our basic three-dimensional unit of volume, we assign it the numerical value of 1. In order to understand how this works, we will take a closer look at what happens when two or more sub-atomic particles combine.

PART II: SOLVING THE MYSTERIES OF QUANTUM REALITY   49

The Primary Level of Symmetric Stability – Quarks in True Units

With the appropriate *integer* values of X_1, X_2, X_3, and Z, in **TRUE** units, the conveyance equation represents the stable combination of three quarks to form a Proton or Neutron. There are many integer solutions for this equation. The smallest integer solution of this Conveyance Equation is $3^3 + 4^3 + 5^3 = 6^3$.

TABLE TWO:

Trial Combination of Two Up-Quarks and One Down-Quark, i.e. The Proton, With Minimal TRUE

Particle	Charge*	Mass/Energy	λ	Total TRUE	MREV**
u₁	+2	4	-1	3	27
u₂	+2	4	0	4	64
d	-1	9	-4	5	125
Total	+3	17	-5	12	216=6³

* For consistency in a quantized reality, charge has also been normalized in these tables.
** **MREV** = Minimum Quantum Rotational Equivalence Volume.

Attempting to use the smallest integer solution of the conveyance equation, $3^3 + 4^3 + 5^3 = 6^3$, to find the appropriate values of λ for the Proton, we obtain negative values for λ for the first up-quark and the down-quark and zero for the second up-quark. It is conceivable that some quarks may contain no λ units, but negative values are not possible because a negative number of total λ units would produce an entity with fewer total TRUE units than the sum of mass/energy units of that entity, violating the conservation of mass and energy. When we try to use the smallest integer solution of the conveyance equation to describe the combination of one up-quark and two down-quarks in a neutron, all of the quarks have negative ג units. See the table below:

TABLE THREE
Trial Combination of One Up-Quark and Two Down-Quarks in TRUE

Particle	Charge	Mass/Energy	ג	Total TRUE	MREV
u	+2	4	-1	3	27
d_1	-1	9	-5	4	64
d_2	-1	9	-4	5	125
Totals	0	22	-10	12	$216=6^3$

In conformance with the Bohr's solution of the EPR paradox (the Copenhagen interpretation of quantum mechanics), newly formed elementary entities do not exist as localized particles in 3S-1T until a 3S-1t measurement or observation is made. This is possible if *all* TRUE units are ג units, undetectable in 3S-1t, before observation and measurement.

The redistribution of TRUE units cannot result in the appearance of negative ג units in the internal structure of a proton because mass/energy and gimmel TRUE units are interchangeable at the sub-quantal level, and the deficit represented by negative ג units would be cancelled by the transfer of mass/energy units to gimmel units. Thus, a negative number of ג units would destroy the particle's identity. Analogous to the axiom 'nature abhors a vacuum', a result of the second law of thermodynamics, some of the TRUE units of mass/energy would fill the deficit of the negative ג units, and the measurable mass/

energy of the particle would no longer be that of a proton and conservation of mass/energy in 3S-1t would be violated. The smallest integer solution of the Conveyance Equation that produces no negative values of ג for the Proton is $6^3 + 8^3 + 10^3 = 12^3$, using this solution we have the electrically and symmetrically stable Proton:

TABLE FOUR: THE PROTON (P⁺)

Particle*	Charge	Mass/Energy	ג	Total TRUE	MREV
u₁	+2	4	2	6	216
u₂	+2	4	4	8	512
d₁	-1	9	1	10	1,000
Total	+3	17	7	24	1728 = 12³

* u1 and u2 have the same number of TRUE of mass and energy, and therefore will register as up-quarks in the collider data but have different numbers of TRUE units of equivalent volume participating as ג to produce the volumetrically symmetric, and therefore stable, Proton.

To see how other structures arise from the combination of quarks, protons and electrons, we need to know how protons, neutrons and electrons relate to the Conveyance Equation: $(X_1)^3 + (X_2)^3 + (X_3)^3 = Z^3$. If the number of TRUE units in the proton is equal to the integer X_1, the number of TRUE units in the neutron = X_2, the number of TRUE units in the electron = X_3, then the resulting compound entity, will be stable in the 3S-1T domain of physical observations.

We have determined that the Proton contains a total of 24 TRUE units of mass/energy and gimmel. We also know that a proton combines with an electron to form a Hydrogen atom, and with additional protons, neutrons and electrons to form all of the stable elements of the Periodic Table. In order to find the smallest solutions of the conveyance equation that are needed to describe the combinations of protons, neutrons and electrons that make up the elements of the universe that we can observe and measure, we must also determine how many TRUE units of mass/energy and gimmel make up the neutron and the electron. To determine how many TRUE units are in a neutron, we can use the same method we used for the proton above.

Up to this point we've used the first two of the smallest primitive integer solutions of the equation $(X_1)^3 + (X_2)^3 + (X_3)^3 = Z^3$: $3^3 + 4^3 + 5^3 = 6^3$ and $1^3 + 6^3 + 8^3 = 9^3$. (A primitive Diophantine solution is defined as one without a common factor in all terms.) We have also used $6^3 + 8^3 + 10^3 = 12^3$, obtained by multiplying the terms of the smallest primitive solution by **2**. The first **12** primitive solutions of $(X_1)^3 + (X_2)^3 + (X_3)^3 = Z^3$ are listed below:

$3^3 + 4^3 + 5^3 = 6^3$	$1^3 + 6^3 + 8^3 = 9^3$;	$3^3 + 10^3 + 18^3 = 19^3$;
$7^3 + 14^3 + 17^3 = 20^3$	$4^3 + 17^3 + 22^3 = 25^3$	$18^3 + 19^3 + 21^3 = 28^3$
$11^3 + 15^3 + 27^3 = 29^3$	$2^3 + 17^3 + 40^3 = 41^3$	$6^3 + 32^3 + 33^3 = 41^3$
$16^3 + 23^3 + 41^3 = 44^3$	$29^3 + 34^3 + 44^3 = 53^3$	$12^3 + 19^3 + 53^3 = 54^3$

As with the proton, the TRUE values making up the neutron must be the smallest possible nonnegative integers, and the sum of the three totals cubed must equal an integer cubed. Thus, we can calculate the number of λ units involved, and the totals of TRUE units required by the conveyance equation to yield results consistent with empirical particle collider data.

The correct unique solution can be found for each triadic subatomic particle by starting with the smallest integer solution of the conveyance equation and moving up the scale until no negative values are obtained. Using the solution $6^3 + 8^3 + 10^3 = 12^3$, the first trial TRUE configuration of the neutron is shown below:

TABLE FIVE:
TRIAL COMBINATION OF ONE UP-QUARK AND TWO DOWN-QUARKS

Particle	Charge	Mass/Energy	λ	Total TRUE Units	MREV
u	+2	4	2	6	216
d₁	-1	9	-1	8	512
d₂	-1	9	1	10	1000
Totals	0	22	2	24	1728=12³

Since this solution still produces a negative value of λ for d_1, we must move to the next larger solution to represent the Neutron: $9^3 + 12^3 + 15^3 = 18^3$

TABLE SIX:
Second Trial of Quark Combinations for the Neutron

Particle	Charge	Mass/Energy	λ	Total TRUE Units	MREV
u₃	+2	4	5	9	729
d₂	-1	9	3	12	1,728
d₃	-1	9	6	15	3,375
Totals	0	22	14	36	5,832=18³

This solution gives us a stable neutron; but we have another problem: None of the solutions with a term equal to 24, the TRUE value of the proton, have a term equal to 36. Nor do any of the solutions listed have two terms with the ratio 24/36. This is a problem because no Diophantine solution means that combinations with equal numbers of protons and neutrons could not be stable, and we know that Hydrogen, Helium, and other elements are stable combinations with equal numbers of protons and neutrons. Looking at the TRUE analysis of Helium for example, we have:

TABLE SEVEN: Attempt to Construct a Helium Atom with $P^+ = 24$ and $N^0 = 36$

Particle	Charge	Mass/Energy	λ	Total TRUE	MREV
2e	-6	2	78	80*	512,000
2P⁺	+6	34	14	48	110,592
2N₀	0	44	28	72	373,248
Totals	0	80	120	200	995,840=(99.861...)³

*Note: The actual number of TRUE units making up the electron is unknown at this point, but this value was chosen because it is the integer value that, with 24 and 36 TRUE units for the proton and neutron, produced a total MREV nearest to an integer cubed, which it must be for a stable Helium atom.

Since a neutron of 36 TRUE units produces an unstable Helium atom, contradicting the empirical fact that stable Helium atoms exist, we have to seek another integer solution of the conveyance equation for the neutron.

Going back to the list of conveyance equation solutions, we see that the next smallest solution that does not generate negatives for the neutron is the primitive solution $7^3 + 14^3 + 17^3 = 20^3$.

TABLE EIGHT:
Third Trial of Quark Combinations for the Neutron

Particle	Charge	Mass/Energy	λ	Total TRUE	MREV
u3	+2	4	3	7	343
d2	-1	9	5	14	2,744
d3	-1	9	8	17	4,913
Totals	0	22	16	38	8,000=20³

This gives us a symmetrically stable Neutron. Next, we need to determine the number of TRUE units for the electron and see if this quark combination for the neutron combined with protons and electrons will yield stable atomic structures. Using the values we have derived for **P⁺** and **N⁰**, i.e., **24** and **38**, the smallest integer solution of the conveyance equation containing the values $X_1 = 24$ and $X_2 = 38$ is obtained by multiplying both sides of the primitive solution $12^3 + 19^3 + 53^3 =$

54^3 by 2, yielding the integer solution $24^3 + 38^3 + 106^3 = 108^3$. Returning to the Helium atom with these TRUE values for **P⁺** and **N⁰**, we have:

TABLE NINE: Helium Atom with $P^+ = 24$ and $N^0 = 38$

Particle	Charge	Mass/Energy	λ	Total TRUE	MREV
2e	-6	2	210	212*	9,528,128
2P⁺	+6	34	14	48	110,592
2N₀	0	44	32	76	438,976
Totals	0	80	256	336	10,077,696=216³

Protons and neutrons contain **24** and **38** TRUE units, respectively, we see that the electron must contain **106** TRUE units for the Helium atom to be stable.

Besides the TRUE that appear as mass/energy in given elementary particles, because of the embedded nature (dimensional tethering) of dimensional domains in TDVP, there must be a minimum number of λ units associated with each particle for stability. Consistent with up- and down-quark decay from the strange quark, the stabilization requirement of an integer solution for the conveyance equation, and the additional TRUE units of λ needed for particle stability, the following table describes the electron, proton and neutron, with up quarks composed of a total of 24 TRUE units, down quarks composed of a total of 38 TRUE units and electrons composed of a total of 106 TRUE units. It therefore represents the normalized mass/energy and λ minimums and total volumes for stable electrons, protons and neutrons, the building blocks of the physical universe.

TABLE TEN: The Building Blocks of the Elements in TRUE

Particle	Charge	Mass/Energy	ג	Total TRUE	MREV
2e	- 6	2	210	212*	9,528,128
2P⁺	+ 6	34	14	48	110,592
2N₀	0	44	32	76	438,976
Totals	0	80	256	336	10,077,696=216³

* Upon measurement, each TRUE occupies the same volume, i.e. the minimal volume for an elementary particle as a spinning object, as required by relativity and defined in TDVP as the basic quantum unit of volume. Each TRUE unit, whether mass/energy or gimmel, contributes to the physical form of a particle, according to the logical pattern in the substrate reflected in the Conveyance Equation, and the relative volume of each particle (in the three dimensions of space) is equal to the total number of TRUE cubed times the shape factor. As noted before, the shape factor cancels out in the Conveyance Equation. For this reason, the right-hand column in these tables contains cubed integers representing the **Relative Equivalence Volume** for each particle making up the combination of atomic and sub-atomic particles.

**The TRUE values for the elementary particles are uniquely determined by conditions necessary for a stable universe. The values for up- and down-quarks are the necessary values for the proton and neutron, as determined above, and the number of ג units and the total TRUE units for the electron are determined by calculating the ג units necessary to form a stable Helium atom. They also determine the smallest possible stable atoms, Hydrogen, Deuterium and Tritium.

It should be clear at this point, that the complex structures of the physical universe do not arise from random particle encounters. Rather, compound structures are formed within the mathematical organization of the Conveyance Equation. With the addition of gimmel, the building blocks of the physical universe have a level of stability and complexity sufficient to form structures capable of supporting life.

The Symmetric Stability of Atoms

Atoms are semi-stable structures composed of electrons, protons and neutrons. They are not as stable as protons, but they are generally more stable than molecules. The Hydrogen atom is unique among the natural elements in that it has only two mass/energy components, the electron and proton. Thus, because Fermat's Last Theorem prohibits the symmetrical combination of two symmetrical particles; they cannot combine to form stable structures like the combination of quarks to form the proton and neutron. The electron, with a small fraction of the mass of the proton, is drawn by electric charge to whirl around the proton, seeking stability. *This means that the Hydrogen atom, the elemental building block of the universe, composed only of the mass and energy of an electron and a proton should be inherently unstable.* So why is it that we have any stable structures at all; why is there a universe? As Leibniz queried: *"why is there something rather than nothing"*?

One of the X_n integers must be 24 to represent the TRUE unit value of the proton, and among the integer solutions of the **m = n = 3** conveyance equation listed above there are four solutions with 24 as one of the X_n solution integers. Nature is parsimonious, and we must never make a mathematical description or demonstration any more complicated than it has to be. Therefore, we start with the smallest solution with 24 as one of the X_n integers. It is $\mathbf{3^3 + 18^3 + 24^3 = 27^3}$. But it does not contain an X_n equal to 38, so we must continue, searching for an integer solution that contains both 24 and 38 on the left side of the equation. Since there are no smaller integer solutions with co-multiples of 24 and 38 as terms in the left side of the equation, we can use the solution that provided a stable

PART II: SOLVING THE MYSTERIES OF QUANTUM REALITY 59

Helium atom: $24^3 + 38^3 + 106^3 = 108^3$. Using it to represent the Hydrogen atom, we have:

TRUE-Unit Analysis for Hydrogen 1 (Protium)

Particle	Charge	Mass/Energy	ג	Total TRUE	Volume
e	-3	1	105	106	1,191,016
P+	+3	17	7	24	13,824
C₂*	0	0	38	38	54,872
Totals	0	18	150	168	1,259,712=108³

* Since the Proton required 17 mass/energy units and 7 ג units, adding up to 24 Total TRUE units, to achieve triadic stability (see the Tables describing the Proton), to achieve the same level of stability as the proton and neutron, the Hydrogen atom must have a third additive component, C₂, consisting of 38 ג units, the third form of the 'stuff' of reality, not measurable as mass or energy in 3S-1t. This satisfies the conveyance equation and produces a stable Hydrogen atom with a total volume of 108^3.

Without the ג units needed by Hydrogen to achieve stability, we would have no universe. The TRUE units of two symmetrically stable entities, the electron and proton, could not combine to form a third symmetrically stable entity (Fermat's Last Theorem). Because of the asymmetry of their form as two symmetric entities of different sizes in TRUE units, they could not combine; they would spiral and be easily separated by any external force. Even if they could adhere to other particles, the resulting universe would be very boring. All multiples of such a building block would have the same chemical characteristics. With the input of the appropriate number of ג units, Hydrogen is a basic building block of symmetrically stable forms in the 3S–1t observable domain of the physical universe.

In 3S-1t, TRUE units can manifest as mass, energy or ג, in order to form symmetrically stable particles and the 168 total

TRUE units of the Hydrogen atom may be arranged in another stable structural form, observed as the simple combination of one electron, one proton and one neutron, known as Deuterium, an *isotope* of Hydrogen (an atom with the same chemical properties).

Hydrogen 2 (Deuterium)

Particle	Charge	Mass/Energy	λ	Total TRUE	Volume
e	-3	1	105	106	1,191,016
P⁺	+3	17	7	24	13,824
N₀	0	22	16	38	54,872
Totals	0	40	128	168	$(108)^3$

Hydrogen 2 (H2) is held together by electrical charge and 128 λ units, 22 less than the H1 atom. This means that H2 is not as stable as H1. What about other isotopes of H1? Is it possible that the TRUE units of a Hydrogen atom or a Deuterium atom can combine with one or more additional neutrons to form semi-stable isotopes? Hydrogen 3 (H3), known as Tritium, is a second isotope of Hydrogen. Its form in TRUE units is represented below.

Hydrogen 3 (Tritium), Valence

Particle	Charge	Mass/Energy	λ	Total TRUE Units	Volume
e	-3	1	105	106	1,191,016
P⁺	+3	17	7	24	13,824
2N⁰	0	44	32	76	438,976
Totals	0	62	144	206	$(118.018...)^3$ *

We see that H3 is an asymmetric structure. One electron, one proton and two neutrons, brought together by attractive forces, cannot combine volumetrically to form a symmetrically stable structure, and as a result, it is unstable and there are very few H3 atoms. Looking at the TRUE structure for H1, H2 and H3, we see that all three are bonded by electrical charge, but H1 has volumetric stability and 150 λ units holding it together; H2 has volumetric stability, more mass/energy units and fewer λ units than H1; and H3 has more mass/energy units and λ units, but no volumetric stability. This explains why H1 is the most abundant, H2 less abundant, and H3 correspondingly less stable. The atomic weights of the elements of the periodic table, in **amu** (atomic mass units), are actually the mean values of atomic masses calculated from a great number of samples. The accepted mean atomic weight for Hydrogen to four significant figures is 1.008. This includes H1 and all isotopes of Hydrogen. If all hydrogen atoms were H1 atoms, this number would be exactly 1. H1 is by far the most stable, and therefore, most abundant, of the Hydrogen family, making up more than 99.99% of all Hydrogen in the universe. Other H isotopes make up the remaining 0.01%, mostly H2, with H3 and other isotopes heavier than H2 occurring only rarely in trace amounts.

The Elements of the Periodic Table

New elements are formed using unique new combination of TRUE, constructed using multiples of the basic building blocks of electrons, protons and neutrons. After Hydrogen came Helium. (See Table Nine above), a stable inert element. The next element is the combination of the inert atom, Helium, with the asymmetric tritium atom, H3 with to form Lithium.

LITHIUM

Particle	Charge	Mass/Energy	ג	Total TRUE	Volume
3e	-9	3	315	318	32,157,432
3P⁺	+9	51	21	72	373,248
4N₀	0	88	64	152	3,511,808
Totals	0	142	400	542	$(330.32...)^3$ *

* Since the total volume is not an integer cubed, Lithium, like Tritium, is asymmetric. It has a stronger electrical bond than H3 and more mass for added stability, but it is less stable because it is asymmetric, and gimmel is only 74% of total TRUE, compared to 89% in hydrogen.

As we proceed with the TRUE analysis, we see that the other elements and compounds necessary for life and the manifestation of consciousness in sentient beings are produced in abundance by the organizing action of the *third form* as ג units, and the conveyance equation.

CARBON

Particle	Charge	Mass/Energy	ג	Total TRUE	MREV
6e	-18	6	630	636	257,259,456
6P⁺	+18	102	42	144	2,985,984
6N₀	0	132	96	228	11,852,352
Totals	0	140	768	1,008	272,097,792=648³

The volume of the carbon atom, $648^3 = (6 \times 108)^3$

NITROGEN

Particle	Charge	Energy/Mass	λ	Total TRUE	MREV
7e	−21	7	735	742	408,518,488
7P⁺	+21	119	49	168	4,741,632
7N₀	0	154	112	266	18,821,096
Totals	0	280	896	1,176	432,081,216 =756³

The volume of the nitrogen atom, $756^3 = (7 \times 108)^3$

OXYGEN

Particle	Charge	Mass/Energy	λ	Total TRUE	MREV
8e	−24	8	840	848	609,800,192
8P⁺	+24	136	56	192	7,077,888
8N₀	0	176	128	304	28,094,464
Totals	0	320	1,024	1,344	644,972,544=864³

The volume of the oxygen atom, $864^3 = (8 \times 108)^3$

The Cube and the Quantum World

Applying the process of rotation and unitary projection from dimension to dimension in space, time and consciousness, we find that the mathematical structure of basic number theory requires the existence of nine finite orthogonal dimensions embedded successively in an infinitely continuous substrate.

Applying the logic of the Calculus of Dimensional Distinctions, an application and extension of George Spencer Brown's Laws of Form, to LHC particle-collider mass/energy data for electrons, protons and neutrons, considered as spinning distinctions of content occupying unitary distinctions of extent in the 3S-1t dimensional domain of physical observations, we find that the light-speed limitation of Einstein's special relativity and Planck's quantization of mass and energy define a minimal unitary distinction, the Triadic Rotational Unit of Equivalence (TRUE). This minimal mass/energy, space-time distinction is the smallest possible finite building block of the 3S-1t universe. As such, it replaces the infinitesimal of the differential calculus of Newton and Leibniz in the mathematical analysis of physical reality. The Calculus of Dimensional Distinctions provides us with the tools needed to continue and extend the work of Minkowski, Einstein, Kaluza, Klein, Pauli, and others who have attempted to use multi-dimensional analysis to integrate and explain the laws of physics.

The process of rotation and unitary orthogonal projection from the planes of one dimension to the next utilizes the Pythagorean Theorem. Generalization of the Pythagorean Theorem equation to three dimensions and application to the minimal quantized distinctions of extent and content produces a set of Diophantine expressions that perfectly describe the combination of elementary particles. Integer solutions of these equations represent stable, symmetric combinations of elementary particles; but when there are no integer solutions, the expressions are inequalities representing unstable combinations that decay quickly.

Fermat's Last Theorem applied to the equation describing the combination of two elementary particles tells us that there are no integer solutions, and thus no stable combinations of two elementary particles. The equation for the combination of three particles, on the other hand, does have integer solutions. This explains why three quarks, not two, combine to form protons and neutrons.

Application of the equation describing the combination of three particles to particle-collider mass/energy data expressed as multiples of the minimal unit, reveals that, in order for stable combinations to form, in addition to the volumetrically equivalent forms of mass and energy, there has to be a third equivalent form that does not register in physical measurements as mass or energy. Representing the third equivalent form as a third variable, gimmel in the equations describing the combination of three particles as integer multiples of the Triadic Rotational Unit of Equivalence (TRUE), we are able to calculate the unique number of mass/energy units and gimmel units needed to produce the stable forms that make up the physical universe, i.e., the elements of the Periodic Table.

Analyzing the new information provided by the third form of the *essence* of the physical universe, we find interesting patterns in the structure of the Elements. For example, Carbon, Nitrogen, Oxygen, and Sulfur have the exact same percentage of gimmel units. This exact ratio in elements that play a major role in life-supporting organisms is not accidental. Without the presence of gimmel units, no stable structures could form and there would be no physical universe. This means that gimmel had to be present from the formation of the very first elementary particle coming out of any big-bang type explosion, guiding the formation of the physical universe to produce structures capable of supporting life. This supports the hypothesis that logical structure, meaning, purpose and life are not emergent epiphenomena, but intrinsic features of reality.

TDVP provides a logical "mechanism" explaining why there is something rather than nothing. In TDVP, the form and structure of reality is determined by the intrinsic logic of nine-dimensional reality, without requiring any transfer of mass or energy. See Close and Neppe 2015 References for the mathematical derivation of TRUE

These results strongly suggest that the stable structures of the physical universe are purposefully formed for use as vehicles through which Primary Consciousness experiences the universe as individual conscious entities. This supports the idea that the physical universe came into existence by intelligent design, not an accident of random combination.

The Origin of Mass

Recall that Max Planck said, "There is no matter as such", that quantum equivalence units (TRUE) are measures of mass, energy and gimmel, and the basic quantum unit of mass is defined as equal to the mass of the electron. But if there is no matter, what is mass?

Hypothesis: *Mass is nothing more and nothing less than resistance to acceleration due to the angular momentum from moments of inertia of the rapidly spinning elementary particles that, in combination, make up an object.* (Acceleration is any change in motion, and angular momentum and moment of inertia are defined as in classical mechanics and relativity.)

The Mass of the Electron, Up-Quarks and Down-Quarks

We have set the mass of the electron equal to unity and determined the masses of up- and down-quarks from collider data (Table One), and we can also determine their inertial masses by applying simple physical principles. For spinning objects, the moment of Inertia is $I = kmr^2$, where m is mass, r is the radius of rotation and the factor k depends on the axis of rotation and the shape of the spinning object. Because we defined the True unit of extent as the radius of the electron, and the mass of the electron as the True unit of mass in our quantum calculus, $I_e = m_e = r_e = 1$. And, therefore, $I_e = km_e r_e^2 \rightarrow k = 1$. This implies that the spinning electron is not spherical, since the moment of inertia of a sphere is $2/5 mr^2$,

There are two possibilities for the shape of the spinning electron: (1.) The electron is shaped like a ring or a hoop spinning about its center of symmetry with $k = 1$, or (2.) the electron is of an unknown shape spinning simultaneously in more than two planes in such a way that the sum of the inertia produced by the spinning planes is 1, and the sum of the k values is 1. In either case, the factor $k = 1$, and $I_e = km_e r_e^2 = 1 \times 1 \times 1^2 = 1$ True unit of mass, indicating that the inertia of a free spinning electron is equal to its mass. In the quantum mathematics of the CoDD, the mass of any free spinning particle is a multiple of m_e, so the next larger spinning particle with a radius, r_u, of $2r_e$ is equal to $I_u = m_e r_u^2 = 1 \times 2^2 = 4$ TRUE of mass, which confirms the mass value of the up-quark. For the next larger particle, with a radius of 3 electron radii, $I_d = m_e r_d^2 = 1 \times 3^2 = 9$ TRUE of mass. These mass values for the electron, up-quark and down-quark agree exactly with the normalized LHC experimental data. Thus, we have shown that for the electron, the up-quark and the down-quark, mass = inertia, proving the hypothesis that, at least for these elementary particles, **mass is equal to the inertia created by spin**. For compound entities, like protons, neutrons and atoms, however, things are a little more complex.

The Mass of the Proton

The proton is a compound entity containing two up-quarks and one down-quark, rapidly spinning as described above. If in combination, the masses of quarks were additive, like apples being added to a basket, the mass of the proton would simply be the masses of the up-quarks and down-quark added together, and the proton should have a mass of 2x4 + 9, or 17 TRUE. But, if mass is the sum of moments of inertia of spinning particles, this will not be the case. The combination will be more like applesauce than apples. The inertial mass contributed by a quark will be greater than its mass as a lone particle because its radius of rotation will be greater. It will be spinning around the center of the compound particle with a larger radius of rotation, and thus the inertial mass of a quark in combination in a proton will be greater.

Assuming similarity of shape for all symmetrically spinning particles, the planes of spin will be mutually perpendicular. The factor **k** for the proton will reflect the fact that it is spinning in three planes in an observational space-time domain of 4 dimensions. So $k_p = 3/4$. Note that, even though we have developed a quantum calculus to apply to quantum reality, the appearance of non-integer values in equations is not a problem as long as the values of measurable quantities of mass, energy and gimmel are integers.

The total units of mass rotating around the center of the proton will be equal to the sum of the units of mass in the constituent particles. So, for the proton, the total mass of the constituents, two up-quarks and one down-quark is **2x4 + 9 = 17**.

Thus, the mass of the proton is $\mathbf{m_p = I_p = k_p(2m_u + m_d)xr_p^2}$ = 3/4(2x4 + 9)(12)² = 3/4x17x(12)² = **1836 TRUE**. This agrees precisely with particle physics experimental data that puts the mass of the proton at 938.27 Mev/c² which converted to normalized quantum equivalence units is 938.27 divided by 0.511 = **1836 TRUE**.

The Mass of the Neutron

The mass of the neutron cannot be determined directly the way we did for the proton for a number of reasons, including its instability as a stand-alone particle. But it has been determined indirectly by subtracting the mass of the protons from the mass of nucleons like the nucleus of the deuterium atom leaving the mass of the neutron plus the binding energy, which can be directly determined. In this way, physicists have determined the mass of the neutron to be 939.5656 MeV, which is equivalent to **1839** TRUE, just 3 units more massive than the proton.

We can't use the same inertial mass equation we used for the proton to determine the effective mass of the neutron because the neutron is formed in a completely different way involving two Hydrogen atoms. But the mass of the neutron *can* be determined using True unit analysis. Like the Hydrogen atom, the way the neutron is formed is unique. The stable Hydrogen atom is formed by the combination of True quantum units of mass, energy and gimmel in accordance with the Diophantine combination equations, called Conveyance Equations because they convey symmetry and mathematical logic into the space-time domain of physical observation. The neutron is formed naturally in the entropic process of two hydrogen atoms forming a deuterium atom, one of the most stable compound structures in the universe. A neutron, if separated from the deuterium atom, decays very quickly; but it becomes an integral part of stable life-supporting atoms when combined with protons and electrons. In the last section of Part II, we will see that TRUE analysis, with gimmel, measurable in quantum units, allows us to calculate the masses of protons and neutrons and account for the "missing mass" in the universe, currently mislabeled as dark matter. To see how non-physical gimmel can affect the total mass of a rotating object, we must use the redefinition of mass as inertia, as described above.

The Origin of the Deuterium Atom and the Neutron

So far, describing reality as consisting of integer combinations of elementary distinctions has been no less reductionist than standard model particle physics. It may even appear that True unit analysis presupposes that reality is built up from the electron as the base unit with 1 unit of mass, incrementally, to more and more complex structures: We will see, however, that this is not the case. The TRUE analysis of particle combinations really starts with the Hydrogen atom.

Hydrogen is the most abundant element in the universe today, and apparently has been very abundant for billions of years, and it is the only atom that contains no neutrons. The Hydrogen atom consists of a proton and an electron. The proton is the stable combination of three quarks, and it is perhaps the most stable known structure, with a half-life longer than the big-bang age of the universe, while the neutron decays in a matter of minutes. According to the standard model of particle

PART II: SOLVING THE MYSTERIES OF QUANTUM REALITY 71

physics, quarks, electrons and neutrinos were the first particles out of the big bang, and within a 100th of a second, quarks began to combine, and about a million years later, atoms began to form [29]. It remains to be seen whether TRUE analysis confirms this picture of the big bang origin of the universe or not. TRUE analysis as applied so far, suggests that simple natural processes going on right now explain the formation of all the elements of the Periodic Table and their isotopes, and they depend on the formation of the neutron in a process involving two Hydrogen atoms.

When one Hydrogen atom encounters another Hydrogen atom, they are electrically neutral, so they don't repel each other, and their two electrons may share a quantized shell surrounding the two protons. This arrangement is asymmetric; but the spinning parts arrive at a more symmetric configuration by ejecting extraneous particles. Table A depicts the 'before' configuration, and Table B depicts the 'after' configuration. In the process, a positron and an electron neutrino are emitted and the very stable Deuterium atom, with an electron, proton and neutron is formed. This process has been observed empirically and is known as "Beta-plus decay".

TABLE A: BEFORE: TWO HYDROGEN ATOMS

2 Hydrogen Atoms	Mass	Gimmel	Total TRUE	Volumetric Equivalence
2e-	2	210	212	2,382,032
2P+	34	14	48	27,648
2C₁	0	76	76	109,744
Totals	36	300	336	2,519,424 = 2x108³

TABLE B: AFTER: DEUTERIUM AND BETA+ EMISSION

Particles	Mass/Energy	Gimmel	Total TRUE	Volumetric Equivalence
e+	1	105	106	1,191,016
v_e	mv_e*	$24 - mv_e$	24	13,824
Energy/Gimmel	-5**	43	38	54,872
Emission Totals	$-4 + mv_e$	$172 - mv_e$	168	$1,109,712 = 108^3$
e-	1	105	106	1,191,016
P+	17	7	24	13,824
N$_0$	$22 - mv_e$	$16 + mv_e$	38	54,872
Deuterium Totals	40	128	168	$1,109,712 = 108^3$
Grand Totals	36	300	336	$2,219,424 = 2 \times 108^3$

*In quantized reality, a particle with no mass or energy/mass equivalence should not exist by itself. In the past, physicists generally considered the mass of the electron neutrino to be at or very near zero. But in 1998, when it was learned that neutrinos oscillate between three types, electron neutrinos, muon neutrinos and tau neutrinos, physicists concluded that neutrinos must have some mass and that it must be less than or equal to a very small, and very specific value. In this representation, the mass/energy of the electron neutrino is unknown, represented by v_e, and v_e mass is estimated at 0.00012 MeV/c², with a confidence level of 95%. Converting this to True unit equivalence, we have: 0.00012/0.511 = 0.00023 times the mass of the electron.

**The negative units in the mass/energy column indicate mass/energy conversion in the entropic decay process as the spinning Hydrogen atoms regain symmetric stability by combining to form a Deuterium atom. This process is known as beta-plus decay.

Comparing the before and after totals in Tables A and B, we see that the process transforms two Hydrogen atoms into one Deuterium atom plus a positron and an electron neutrino; but the total number of TRUE and total volumetric equivalence units in the Deuterium atom plus emissions remain unchanged from the totals before the combination demonstrating conservation of mass, energy and gimmel.

Conservation of mass, energy and gimmel in dynamic systems ensures that the moment of inertia of a spinning particle that becomes part of a compound spinning particle is conserved in the total angular momentum of the larger particle. The moment of inertia of a particle rotating around the center of symmetry of a larger compound particle, multiplied by the radius of revolution gives the contribution of the smaller particle to the total inertia of the compound particle. The total inertia within the compound particle is the measurable mass of the particle.

To find the TRUE inertial mass of the neutron we must use information from Tables A and B. The neutron is different than the proton in three important ways: (1.) the sum of the inertial masses of the quarks involved, (2.) electrical charge or lack of it, and (3.) the way they are formed which may affect the value of the factor **k**. The Hydrogen atom is formed of quantum units of mass, energy and gimmel in accordance with the Diophantine combination equations, while the neutron is formed in situ as two Hydrogen atoms are transformed into a deuterium atom by beta-plus decay.

To determine the inertial mass of the neutron, we must determine how these differences affect the general equation for the inertia of a spinning object: $I = kmr^2$. The neutron's effective mass, m_{vn} is the sum of the masses of the neutron's constituents modified by effective radius and spin applied in the factor **k**. From Table B, the mass of the neutron = $m_u + 2m_d + m_{ve} = 4 + 2 \times 9 + m_{ve} = 22 + m_{ve}$, and we see that 5 units of the 22 were converted from gimmel to mass in the process, while 17 of the quark masses came from one of the combining Hydrogen atoms, so the factor **k** and the radius of rotation for 17 of the neutron's quark units is the same as for the proton. The charge of the neutron is zero because the opposing spins of the quarks cancel; and the opposing spins also cancel the angular momentum causing the 5 quark mass units to collapse to a spherical shape at the center of the neutron structure and contribute directly to the total with an effective radius of 2 linear units. The formula for the inertia of a spherical spinning object is $I = 2/5mr^2$. Applying these factors, the inertial mass of the neutron becomes $m_n = 3/4\{17 \times 12^2 + 2/5(5 + m_{ve}) \times 2\} = \{3/4 \times 17 \times 12^2 + 3/4 \times 2/5 \times (5 + m_{ve}) \times 2\} = 1836 + 3/5(5 + m_{ve}) = 1836 + 3 + 3/5 \times m_{ve}$.

Replacing the unknown m_{ve} in this equation with the maximum possible value of the mass of the electron neutrino from empirical data, we have $m_n = 1836 + 3 + 3/5(0.00023) = 1839 + 0.000138$ which translates to **1839 normalized TRUE** and is in agreement with empirical data.

Using TRUE analysis, we have seen how neutrons and deuterium atoms are formed from hydrogen atoms, and we've calculated the mass of the neutrons formed in this process. As a check on the validity of the TRUE analysis methodology, we can calculate the mass of the neutron from experimental data in another way.

The mass of the Hydrogen atom obtained by adding the mass of the proton and the electron is 1836 + 1 = 1837 TRUE. The mass of deuterium, an isotope of Hydrogen, containing all three elementary particles, has been determined experimentally to be 1878.4 Mev/c². We can convert this to TRUE by dividing by the mass of the electron in Mev/c²: 1878.4 Mev/c² divided by

0.511 Mev/c² = 3676 TRUE. We can now subtract the mass of the proton and electron from the mass of the deuterium isotope to obtain the mass of the neutron. Thus, the mass of the neutron is 3676 − 1837 = 1839 TRUE. So, the masses of the electron, proton and neutron in quantum equivalence units are **1, 1836 and 1839**, respectively.

Converting these masses to Mev/c², the units of mass used in the current paradigm, we have 0.511, 938.2, and 939.7 Mev/c², values that are within the margins of rounding and sampling error of the empirical values obtained experimentally, thus validating the hypothesis that mass is the result of the inertial resistance created by the spinning of elementary particles.

Simulating Atomic, Molecular and Galactic Structures with Cubes

As we saw above, the hydrogen atom is 108 TRUE cubed, consisting of the quantum units (TRUE) of an electron and a proton combined, with mass/energy and gimmel units included. The combination of up-quarks and down-quarks can be visualized as combining three cubes, 24, 38 and 106 TRUE on each side, representing an electron, a proton and gimmel, to produce a cube 108 TRUE on each side, equal to a total of 216 cubes like the Rubik's cube, when each sub cube represents a TRUE volumetric quantum unit. To match experimental observations of hydrogen atoms and normalized data from the Large Hadron Collider, two such combinations are required.

Notice that two 3x3x3 cubes have a total of 54 + 54 = 108 colored faces and that $108 = 1^3 \times 2^3 \times 3^3$. Also, the TRUE quantum units that form up-quarks and down-quarks reflect the spin in three to nine planes of rotation. The normalized electron mass in TRUE units due to planar spin is 1^2, up-quark inertial mass is 2^2 and down-quark inertial mass is 3^2. These patterns provide a glimpse of the remarkable numerical patterns that appear in physical reality from the quantum scale, to the everyday macro-scale, through to the galactic, or cosmic scale, when the mathematics of observation and measurement are normalized to the TRUE quantum unit. As demonstrated in the sections above, this simulation approach can be applied to all of the atoms of the periodic table of elements, and thus to the macro and galactic scales, fully uniting classical physics, quantum physics and relativity, with the primary logic of the Calculus of Dimensional Distinctions.

PART III: SOLVING THE MYSTERIES OF CONSCIOUSNESS AND LIFE

Memory and Intuition

There are several concepts that I want to bring from our contemplations of the cube in Part I, and quantum reality in part II, into Part III, including the Triadic Rotational Units of Equivalence (TRUE) quantum units of mass, energy and *gimmel*, the third form of reality, the universal connectedness of all things, and the importance of memory and intuition.

Without memory and intuition, we are adrift in an infinite sea; the world is like shifting sand beneath our feet. We seem to have been thrust, without being prepared, into an on-going drama, where we must try to survive in a harsh and brutal environment. Looking back, birth may seem like awakening from a deep sleep: Upon awakening, we look around and find that we have arrived in the middle of the strange story of life on this planet, with no knowledge of how it began, and even worse, no clue about what our role might be and how it might end. The whole scenario seems to be a mystery. Our awareness of self seems unconnected with the world we experience. The only thing we know for sure is that we exist - and that we need to know more, in order to continue to exist.

In most cases, we have no conscious memory of what happened before we open our eyes as an infant, but this is an illusion. We have a deep somatic memory hidden in the RNA/DNA structure of our forming bodies. We are not born as empty containers, little blobs of unstructured protoplasm with no knowledge of the past. *Not at all.* In fact, each blob of protoplasm blossoming into this world has within it a vast storehouse of memories, physically manifest in complex single and double spiral structures called RNA and DNA, containing records of the distant past and blueprints for the future, reflecting the incredible narrative of intelligent design.

Also hidden from us, as long as our consciousness is focused on physical reality and attached to the physical body, is the complete record of the process of the manifestation of Primary Consciousness in the finite worlds of nine dimensions. This memory in Primary Consciousness, existing beyond time and space, is called the *Akashic Record*. *Akasha* is a Sanskrit term that has been interpreted by western thinkers as space or aether (also spelled ether). This is a gross oversimplification of the Sanskrit term in my opinion. A more accurate translation would be: Akasha is the *conscious* sub-quantum fabric of reality, consisting of patterns of gimmel.

Our sub-conscious and super-conscious connection with Primary Consciousness is the reason we have an intuitive sense of direction, of meaning and purpose. Without this intuition, we are nothing, full of sound and fury for a while, signifying nothing. Without intuition and the somatic memories of the Akashic record imprinted in our DNA, we have no way of knowing who we are, where we came from, or where we are going. We only know that we are conscious, and truly, for us, our **reality begins and ends with consciousness.**

PART III: SOLVING THE MYSTERIES OF CONSCIOUSNESS AND LIFE 79

Memory in Matter

The configuration of a cube at any given point in time is imprinted in the Akashic record. Therefore, the Akasha contains a record of the moves that brought it to that configuration. That record is the cube's memory of its past. In the same way, the current configuration of matter in the physical universe contains a record of the past, i.e. reality's memory of how it came to be what it is now. Superficial examples of this are: the geologic record, and the light from the distant stars. Records of the floods and droughts of the past, volcanic eruptions, the warming and cooling of the planet, all are recorded in the lithology of the crust of the earth. The light from distant stars gives us a picture in consciousness of what was going on in the cosmos billions of years ago.

Our brains and bodies are composites of memories of events in the past that resulted in the form we now occupy. Memories are stored in the Akasha, and reflected everywhere, in the quarks and stars, in planets and galaxies, and in the DNA and RNA molecules of our bodies, we just need to know how to read them. In finite physical systems like the cube, only a finite number of memories can be stored. However, each one of the 43 quadrillion plus possible configurations is unique, and can be traced back through at least 9x6x2 =108 moves. (9 planes of rotation times 6 cardinal directions, times the parity of clockwise or counter-clockwise rotations.)

The trace of movements in the media of inert matter, however, is meaningless and soon lost if not connected with consciousness as patterns in gimmel. When you are solving the cube, the movements of the planes of rotation are being accomplished intentionally, and there is therefore, some memory of the sequence of the moves in your consciousness. Let's take the most extreme example: Suppose you start with a cube with all six sides monochromatic, i.e. all red on one face, all green on another face, etc. You proceed to scramble it, *and* in the process, you memorize each and every move. Then, you can "solve" the cube by just reversing the moves that scrambled it to arrive back at the coherent or solved state.

If memorizing a great number of rotations, some clockwise and some counter-clockwise, is too onerous, you can record the scrambling moves as you make them on a recording, or by writing them on a piece of paper. These would then be a secondary recordings of memories of sequences of events. These recordings, however, like the lithological records in stone and the vibrational records in light, are symbolic and mute without the interaction of consciousness. Until they are interpreted and applied by a conscious entity, they are meaningless data. Memory in the physical cube itself is limited to the record of the structure of its form in the molecules of plastic, the atoms, electrons, quarks, and TRUE units of mass, energy and gimmel that make up its structure.

Notice how, in the process of solving the cube, just as in quantum physics and in the course of our lives, there is no one simple formula with a set number of rules and deterministic steps to the solution. There are many choices at every turn, and many, if not most of them may lead into time-consuming loops that take us nowhere, and often only make matters worse. In life, as in the course of solving the cube, the path from pain, chaos and confusion to a coherent and stable existence is neither easy nor obvious. We are constantly struggling to understand where we are in the process of learning the lessons of life, looking for hints, in the form of patterns that occur as results of our actions. Sometimes we make wrong decisions that lead to time-consuming sidetracks, over and over, just as can happen in the process of solving the cube. But there is always something to be learned, if we are alert.

In both cases, solving the cube and living life, some who have gone before us and made some sense of things, have left hints and suggestions, even instructions and guidelines, but such guidelines are often misinterpreted by others with no better understanding than our own. The proverbial blind leading the blind, unfortunately often, both fall into the same ditch!

Even the best of instructions are only guidelines. The map is never identical with the actual territory that we must traverse. Even with the best teachers and guides, we can fail to solve the mysteries of life, because in the final analysis, the results depend upon our own actions, actions that no one can perform for us. However difficult and torturous it may be, we must walk the path to our destiny ourselves.

Analogies and Patterns

With the cube, the solution is achieved by carefully observing the patterns that appear in the arrangement of the colored faces of the three layers and developing our intuition to improve our ability to choose the moves that will lead to symmetric coherence. The three layers of the cube are inter-connected by nine sets of pieces that may be rotated in nine orthogonal planes. The logic and mathematics of nine-dimensional geometry is the same for all manifestations of reality, quantum, personal and cosmic. Consider these analogies:

The first layer, which is the easiest to complete, is analogous to the physical body, and a *completed* first layer is analogous to a strong, healthily structured body. The second layer, analogous to the mind, is a bit more difficult to complete, and a completed second layer is analogous to a clear and healthy mental state. Completion of the third layer, analogous to spirit, while maintaining the coherence of the first two layers, is the most difficult to attain, and is analogous to achieving the ultimate enlightenment, also known as cosmic consciousness. This term was probably first used by Dr, Richard M. Bucke, MD, in his classic book by the same title.

The three layers taken together as a unit, i.e. the cube, is analogous to the individual soul, and the interactions of the layers with each other and the rest of reality through rotations of the nine orthogonal planes, are analogous to the individual soul's interactions with the nine dimensions of reality. A completed cube, with color coherence and the six dimensions aligned with the six directions, is analogous to the state of cosmic consciousness, which is the alignment of all aspects of life, physical, mental and spiritual, with the ultimate state of reality. Finally, the purpose of life is the full realization of the consciousness of the oneness of all things.

Most of us on this planet at this point in time, are in various states of confusion much like the scrambled cube. Each one of us exists in our own unique scrambled state, a reflection of our past choices and actions. The question is: How do we become aligned with the positive patterns of reality, traditionally known as the will of God?

The Elegance of the Three, Six and Nine, and the Cube

"If you only knew the magnificence of the 3, 6 and 9, then you would have a key to the universe." – Nikola Tesla

The cube has **3** dimensions, **6** sides and **9** rotational planes. There are **3** forms of the essence of reality, **6** relationships between them, and each of the three forms have 3 dimensions, making 3x3 = **9** dimensions in total.

There are three worlds existing within the reality we experience. They are not separate. They seem to be separate only because of our limited perceptions of them. The world of the experiences of conscious beings seems to exist suspended between the quantum world of elementary particles, atoms and molecules and the cosmic world of planets, suns and galaxies. However, these three worlds are intimately related, and, as we learn, we are aware of some of the relationships, but we cannot be completely aware of their multi-dimensional connections until we reach the final, coherent state of cosmic consciousness.

Analogous to the progression from one of the scrambled states of the cube to the solved state of color coherence, the purpose of life is to progress from the random confusion of an incoherent body, mind and soul to the beauty, elegance and stability of complete alignment with universal law and cosmic consciousness. This is the magnificence Tesla spoke of. It is reflected in the dimensions of the quantum realm, the macro-world and the cosmos, and it is reflected in the 3 dimensions, 6 directions and 9 rotational planes of the cube. Let's see how it is reflected in life and human consciousness.

Parallel Patterns

We've seen how TRUE, as units of quantum reality, appear as mass, energy and gimmel in elementary particles. Because of the embedded nature of dimensional domains, there must be a specific number of units of gimmel associated with each sub-atomic particle for stability. Consistent with up- and down-quark decay from the strange quark, the quantum requirement of an integer solution for the conveyance equation, and the additional units of gimmel needed for particle stability, every electron, proton and neutron must contain specific numbers of TRUE.

TRUE analysis of the elements of the periodic table shows us that the highest percentages of gimmel are found in the elements that support life, elements like Carbon, Hydrogen, Oxygen, Sulfur, and Nitrogen. In what may seem like a remarkable coincidence, these elements along with the free electrons of electric charge, form the CHOSEN elements of life, and the Hydrogen -Oxygen compound we call water also has one of the highest levels of gimmel among stable molecules. This shows that gimmel is the signature of life and individualized consciousness.

The Vehicle of Consciousness

The most important discovery of TDVP is the discovery of the necessity of the existence of the third finite form of reality, which we are calling gimmel, in addition to mass and energy, gimmel is mathematically and physically required for the sub-atomic, atomic, and molecular stability that supports the organic life forms that function as vehicles for individualized consciousness in order that we, as finite reflections of the logical patterns of Primary Consciousness, may experience reality. Without gimmel, there would be no symmetric stability and no physical universe. Awareness of the distribution of gimmel in atomic structure leads to the realization that organic life is the vehicle of consciousness, which makes the experience of reality possible through the awareness of the "I am" within finite structures. Life, in its many forms, is a finite manifestation of pervasive Primary Consciousness, which is infinite.

Sacred Dimensionometry and Alignment

When we modeled elementary particles with the cube in Part II, we saw that they are triadic, i.e. their substance exists in three forms: mass, energy and gimmel. The composition of conscious entities is based in the same triadic nature of reality as electrons, quarks and atoms, only more complex. Like the electron, up-quark and down-quark, the finite conscious entity is composed of various meaningful configurations of mass, energy and gimmel. Every cell of a living organism consists of combinations of mass, energy and gimmel. None of the structures that make up the physical universe and support life and consciousness would be stable without all three, and no scientific model is complete without all three.

A conscious entity has limited control of the mass and energy of its body, within the limits of the logical structure that exists throughout all dimensional domains of reality, through the action of gimmel, the underlying fabric of the universe. It is in this way that all things in the universe are connected. This is the basis of the statement "As above, so below", and the Biblical statement: "Let us make mankind **in our image**, **in our** likeness... For in **the image of God** He **made** man" - Genesis 1:26-27.

Sacred dimensionometry is the logical space-time-consciousness structure of the nine dimensions of reality. To the extent that the physical structure of the body of a conscious being is aligned with the logically meaningful structure of Primary Consciousness, and that structure is also in alignment with the symmetric framework of the space, time and consciousness domains, the being is healthy in body, mind and spirit, and is capable of being a positive force in the universe.

Without this alignment, the various energies that support the functions of the living organism cannot flow freely from the patterns of gimmel in the primary conscious substrate through the channels of space, time and consciousness into the mind and body of the conscious being. To the extent that this flow is restricted or blocked, the conscious entity may experience physical, mental, and/or spiritual disease and will be incapable of functioning as it should. In this blocked or restricted state, a conscious entity cannot easily manifest, or even begin to comprehend the higher states of consciousness.

We have seen that the cube can be useful as a model of quantum reality, and how the challenges of solving the cube are analogous to the vicissitudes of solving life's problems, including spiritual advancement. In Part IV we will see how solving the mysteries of the cosmological realm is also relatable to the solving of the cube.

PART IV: SOLVING THE MYSTERIES OF THE COSMOS

"Somewhere, something incredible is waiting to be known." - Carl Sagan

Embedded Dimensional Domains

Embedded within the transfinite substrate are three dimensions of space, three dimensions of time and three dimensions of consciousness. These nine-dimensional domains are temporarily contracted to three dimensions: one of space, one of time and one of consciousness during observations and condensed into the distinctions of spinning energy (energy vortices) that form the structure of what we perceive as the physical universe.

In the limited humanly observable domain of 3S-1t, this spectrum ranges from the photon, which is perceived as pure energy, to the electron, with a tiny amount of inertial mass (0.51 $MeV/c^2 \approx 1 \times 10^{-47}$ kg.), which we've normalized to one TRUE quantum unit, to quarks ranging from the "up" quark at about 2.4 MeV/c^2 (normalized to 4 TRUE) to the "top" quark at about 1.7×10^5 MeV/c^2, to the hydrogen atom at about 1×10^9 MeV/c^2 (1.67×10^{-27} kg.), or 108^3 TRUE, to the heaviest known element, Copernicium (named after Nicolaus Copernicus) at 1.86×10^{-24} kg[1]. So, the heaviest atom has about 10^{23} times, that is, about 100,000,000,000,000,000,000,000 or 100 sextillion times heavier than the inertial mass of the lightest particle, the electron.

---

[1] Cn and atomic number 112 was created in 1996. It is an extremely radioactive synthetic element that, as far as we know, has only been created in a laboratory. The most stable known isotope is copernicium-285 (ref Wiki)

All the Elements of the Periodic Table are made up of stable vortical distinctions that are known as fermions, "particles" with an intrinsic angular spin of 1/2, or they are made up of combinations of fermions. **Table One,** above in **Part II**, lists the fermions that make up the hydrogen atom and their parameters of spin, charge and mass based on experimental data.

Metaphysical Implications

We have seen how gimmel, the finite, measurable manifestation of Primary Consciousness in the physical universe, shapes reality from the quantum level to the cosmic level, and how experimental data and the known laws of dimensional mathematics and physics reveal the existence of gimmel and our connection with infinite Primary Consciousness. In the process of learning who and what we really are, we become aware of all our memories and our connection with Primary Consciousness. Achieving this level of individual awareness is called cosmic consciousness. Cosmic consciousness is the purpose of finite reality, the physical universe, life, and the consciousness manifested in it, and the driving force behind all change.

The discovery and proof of the existence of gimmel eliminates materialism as a viable basis for science, philosophy and human understanding in general.

Spiritual evolution is the driving force behind the universe, not some random meaningless explosion, the conclusion of reductionist materialistic science. Reality does not exist and cannot exist without consciousness. The discovery of gimmel is empirical proof of this.

The question of whether reality has existed, or can exist, without consciousness, sounds like a question that science should be able to answer, but, in a logically structured reality, it is in fact, an utterly meaningless question. It certainly is not a scientific question, because a scientific question must be testable. Science, as it currently exists, like an awakening giant, intuitively presupposes the existence of consciousness in the form of conscious beings, thinkers and observers who can play the parts of all sorts of characters in this on-going drama, developing hypotheses about reality. But no hypothesis is possible without conscious thought, and no hypothesis can be tested without a conscious observer. Thus, the question of whether the universe could exist without consciousness is a truly nonsensical question because a universe without an observer cannot be observed.

The Scientific Basis of Infinite 3-D Time

The polymath Gottfried Wilhelm Leibniz (1646 -1716) asked a much more important and meaningful question, one of the most important questions ever asked, a question that he thought science should answer first. That question was: "Why is there something, rather than nothing?" This may appear to be an unanswerable question at first, but until it is answered, science is stumbling in the dark, dreaming up hypothesis after hypothesis based on *a priori* assumptions, physical observations, imagination, and beliefs.

Leibniz's question is not a *meaningless* question because, if there ever was absolute nothingness, and there is no basis for believing that absolute nothingness can spontaneously produce anything, there could never, ever be anything. On the other hand, because there *is* something, while it is continually changing, it will always be around in some form, therefore there has never been a state of absolute nothingness.

Before you dismiss these statements as metaphysical hypotheses that can never be proved or disproved, much like the belief that the universe would exist much as it does with or without consciousness, they *can* be proved, and not only that, the laws of physics as we know them so far, strongly support them, and depend upon them.

In the Beginning …

Where did everything come from? Was the universe as we know it engineered by a conscious intelligence to have purpose and meaning, or did it just happen by accident? Is this a question that can be answered within the scope of human intelligence? Many answers have been offered over thousands of years of human history by thinkers of all sorts: philosophers, theologians, scientists, and mystics. But, are any of the answers truly final and definitive? Or do they come with arguments convincing enough to compel you to live your life as if they were true? Apparently, many people have thought so, because over the history of life on this planet, bloody wars have been fought over some of the answers to this question, and many people have died defending their beliefs in what they considered to be the correct answer to this question.

If you have been reading my books, papers, and/or posts, you probably won't be surprised to learn that I think there *is* a definitive answer to this question. Do I dare offer my answer for you and the rest of the world to consider? Why not? At my age and stage in life, it makes little difference whether anyone listens, agrees, disagrees, or simply ignores me. I am happy with my answer, and I find that that is enough for me. You are free to accept it, reject it, think about it, or ignore it. It's completely up to you.

A Starting Point

I think we can agree that, without question, *there is something real that actually exists, and you and I are part of it.* Without this knowledge of an existing reality containing at least you, me and the universe, we have nothing to talk about. So, given that there *is something*, how did this *something* come to be what it is now? That is the question, and there are three possible answers: 1) Something from nothing, 2) nothing from something, and 3) something from something else.

To believe that number 1 or 2 is the reality, you have to discount nearly all the evidence at hand. No one has ever seen something arising from nothing, or something disappearing into nothing. Even when it appears that way, a thorough investigation reveals that one of the most basic laws of physics holds in every case: the law of the conservation of mass and energy. In all of the experiments ever conducted into the physical, chemical and biological processes of our universe, we see only change, never creation from nothing, nor total annihilation of anything. In even the most violent explosion, the sum total of all matter and energy before and after the explosion is the same. In other words, the empirical evidence from our experience is for number 3, not 2 or 1. Something never arises from nothing, something never degrades to nothing and the something we have now came from something else, because it was different in the past, and, in our dynamic reality, it will be changed from what it is now into something else, but the sum total of mass and energy will remain the same.

Despite the evidence, historically, mainstream science and mainstream religion have both declared that 1, 2 or a combination of them is the true nature of reality. In the theory of a big-bang expanding universe, e.g., the equations of general relativity predict a mathematical singularity at the origin event and mathematical singularities in black holes, with the semi-stable world of our experience existing somewhere in between. The current scientific paradigm sees reality expanding from a mathematical singularity at the beginning of spacetime, with everything eventually falling into the singularities of black holes. This is a process of something from nothing (1) becoming nothing from something (2) unless you assume that the nothing is not really nothing, but some other form of the something we have now; but then, you have number 3, don't you? In the current scientific paradigm, quantum field theory (QFT), with particles defined as quantized states of underlying fields which are more fundamental than the particles, is, in my opinion, closest to reality. But, QFT, using matrices with values subject to the Heisenberg uncertainty principle as perturbations of the underlying fields is more descriptive than it is explanatory.

In Catholic theology, *creatio ex nihilo* (creation out of nothing), is a doctrine invented by early Catholic theologians after the original teachings of the pre-Christian Judeans, Jesus and the first Christian theologian, Origen, were subverted by the Emperor Justinian I in his anathemas against Origen, in 553 AD. Justinian realized that the teachings of the Jewish Gnostics and the followers of Jesus constituted a serious threat to his power because in his interpretation of early Christian teachings, Origen had written:

"Each soul enters the world strengthened by the victories or weakened by the defects of its past lives. Its place in this world is determined by past virtues and shortcomings."

Such teachings were in direct conflict with what Justinian saw as his divine birthright as a Roman Emperor, to rule the world, so he seized on this statement and related ideas in early Christian doctrine, as documented by Origen, that undermined the exclusivity of the Roman Emperors' claim of divinity. If people were allowed to believe that by being virtuous, they could rise to the level of an Emperor, i.e., to the status of a god, or even sons and daughters of God, then the power of the Emperors would be seriously threatened. He decided that he must declare this idea to be heresy and take strong measures to stamp it out. The *anathemas*, an edict that he prepared for this purpose, read in part:

"Whosoever teaches the doctrine of a supposed pre-birth existence of the soul, and speaks of a monstrous restoration of this, is cursed. Such heretics will be executed, their writings burned, and their property will become the property of the Emperor."

This was, of course a powerful incentive for Christians priests and monks to remove any such references from the scriptures from which they taught. Without the teaching of the eternal nature of the soul, theologians were free to shape the doctrine of the church in a way that ensured that the masses had to depend upon the Church of the Holy Roman Empire for salvation. It was a way to control the masses, pure and simple, and perpetuate the power of the Holy Roman Empire. Most other major religions, with the exception of Islam, which, like Judaism and Christianity, is also of Abrahamic origin, are not encumbered by this illogical assumption of *creatio ex nihilo*.

So what does it mean if number 3 is the real answer? It changes things very profoundly: With no absolute beginning or end, we must look at human history in a completely different way. No longer burdened by the misconception that everything was created out of nothing and that consciousness is something emerging from organic neurology evolving only a short time ago, we begin to see that a simplistic linear view of things is very misleading. The cyclic nature of things taught by the philosophies of the orient make more sense.

Researchers who claim there is evidence that some of the ancient stone structures scattered around the planet are much older than mainstream archeologists believe, may not be as wacky as they seem. When viewed through the lens of belief in answers 1 and/or 2, their claims can't be right, but if you drop the irrational belief in a linear progression from nothing to something, and accept the evidence for the eternal existence of everything in some form, you have to take their evidence seriously, because civilization, just like everything else, undoubtedly progress in cycles. Our fixation that we are the epitome of the development of sentient species for all time, due to the illogical belief in answers 1 and 2, is as egocentric and as wrong as the idea widely believed a few hundred years ago that the earth was flat and the center of the universe.

I have found that the logical structure of the universe is reflected in the logical structure of pure mathematics, and vice versa. This finding, combined with recognition of the endless process of something from something else, means that physical reality is a quantized logical structure embedded in the infinitely continuous multi-dimensional field of consciousness, and the illusion of beginnings and ends is only meaningful in relation to the amount that we are identified with finite physical bodies. Identification with the undifferentiated field of consciousness allows us to see time in the same way we see space: three dimensional. Once freed from the illusion of being limited to finite three-dimensional objects evolving in one, unidirectional dimension of time, and rising into the perception of 3-D time, we see that everything exists eternally, and only appears to evolve in cycles of finite duration, due to the physical limitations of human perception.

So, the answer to the original question: *"Where did everything come from?"* is *everything* has always existed. There is no absolute beginning or end, only endless cycles of change. This answers a lot of otherwise unanswerable questions, including Leibniz's question, the first question natural science should answer: "Why is there something rather than nothing?" There is something because there has always been something. Nothingness is an illusion. The illusions of absolute beginnings or ends are perpetuated by certain traumatic changes, like the birth and death of physical bodies. So, everything that is something came from something else that existed before the beginning of the process or processes that changed it into the something we have now, and the something we have now is already being transformed into the something else that will exist in the future. But awareness expanded into the 3-D time and 3-D consciousness predicted by pure mathematics, becomes awareness of the reality behind the illusions of 3 space- 1 time.

In the beginning of one cycle, we find the end of the previous cycle, but they are not the same. The new cycle is one of greater awareness than the previous one, because we have learned and expanded our awareness; and thus we rise in a progressive spiral from the finite into the Infinite.

Let's start with the statement that nothing *cannot* produce something. If you are familiar with quantum field theory, you might argue that virtual particles produced in a vacuum in experiments demonstrating the Casimir effect, prove that something *can* come from nothing. In fact, they prove just the opposite. They prove that what we define as a vacuum is *not* a state of absolute nothingness. Experimental data prove that it is filled with the quantum field, and all particles are excitations of the quantum field. Experimental data have shown that such virtual particles do not violate the conservation of mass and energy, as they would if they were something from nothing.

You might think that the big bang is the prime example of something from nothing. It is not. Most theoretical physicists have backed off of the mathematical singularity theory that Hawking and Penrose thought they had proved several years ago, and believe that all of the mass and energy of the universe existed in a tiny, extremely hot mass before the big bang, and the law of conservation of mass and energy, as well as the second law of thermodynamics, the law of increasing entropy in finite physical systems, appear to hold throughout the known history of the universe, and there is no evidence to the contrary.

The law of the conservation of mass and energy says that nothing is ever created or destroyed, it only changes form. And the second law of thermodynamics says that, in an expanding universe, complex structure always tends to break down, and never becomes more complex except in the case where living organisms borrow from their surrounding environment to build stable complex non-entropic structures. This activity of living organisms increases ordered structure locally, but reduces it in surrounding areas. This accounts for organic complexity, without violating the second law of thermodynamics because the overall complexity continues to decrease, or as physicists say, entropy increases over all.

Living organisms do this kind of borrowing, but if life and consciousness did not exist in the universe until a long time after the big bang, as is assumed in the current paradigm, the universe could never be more structured than it was immediately after the big bang. This, and our findings to date, suggest that there can be no absolute beginning or end of the physical universe, only change.

In addition to this argument, there is mathematical proof based on a huge amount of empirical data, that there is something rather than nothing. The proof has already been published but a brief summary of it is warranted here for the reader who is not interested in seeing all of the mathematical detail. Before the proof can be developed, we have to clear up some misconceptions about, and misuses of mathematical logic.

Symbolic Representation of the Structure of Reality

The Greek philosophers Pythagoras and Euclid developed much of the logical and geometrical basis for the mathematics of modern science. The axiomatic concepts upon which their work was based were drawn from observations of the natural world. They were, in fact, symbolic representations of the structure of reality. But the various forms of mathematical methods based on their work developed over the past few hundred years, have been used primarily for practical engineering and technical purposes.

Because of this, most scientists, and virtually all engineers and technicians, consider mathematics to be nothing more than a tool for solving problems, and the connection between mathematical logic and reality has been blurred, if not completely lost. Because of this loss of connection with the metaphysics of reality, we find modern scientists and mathematicians expressing surprise when theorems developed in pure mathematics turn out to have direct correspondence with real, observable physical phenomena.

This loss of awareness of the connection between mathematics and reality is the hidden cause of a major problem in modern scientific efforts to produce a complete successful mathematical representation of reality. Without re-establishing this most basic connection between mathematics and reality, finding anything approaching a "theory of everything" is impossible. This problem, endemic in modern science, once identified, is surprisingly easy to resolve. Let's have a brief look at how the problem arises, and the solution.

The Limitation of Physical Division

Some statements we hear from modern scientists are symptomatic of the problem: Physicists today often declare that there are two different sets of mathematical rules for physical reality, one for the macroscale and another strange, counter-intuitive set of rules for the quantum realm. And they can't seem to resolve the conflicts between the two sets of rules. Intuitively, we know that reality is internally consistent, it is the standard model that is incomplete. The conflicts between relativity and quantum theory really just indicate that one or both sets of rules is either partially wrong or incomplete.

At the root of the problem is the misapplication of a mathematical method that has been known as "the calculus" for more than 300 years: the calculus of Newton and Leibniz. The fact that it is called *the calculus*, and not just "a calculus", is also a symptom of the problem. It is in fact, just one of several calculi that can be defined at the interfaces of dimensional domains. It happens to be defined at the interface of three dimensions of space and one dimension of time, which makes it very useful for analyzing and arriving at sufficiently accurate results for solving problems involving motion on the macro-scale.

Like all puzzles, once the nature of the problem is understood, and a solution is found, we wonder how we could have missed it for so long. This is no different: The fact that "the calculus" doesn't apply at the quantum level should have been obvious to anyone familiar with the theoretical basis of Newtonian calculus. The only excuse is that the calculus was so successful solving everyday macro-scale problems that we simply didn't think to re-think the basic assumptions behind it. If we had, it would have been obvious that there was a serious problem.

Newton's calculus depends on the assumption that equations describing physical processes are continuous functions with variables that can approach zero infinitely closely. Quantum reality, on the other hand, is not continuous, and the variables describing it are not infinitely divisible.

Put as simply as possible: There is a bottom to physical reality. The structures of physical reality cannot be divided indefinitely. This means that the calculus of Newton and Leibniz, while very useful at the mid-scale of reality, is inappropriate for application at the quantum level.

This problem with Newtonian calculus has no effect on most macro-scale computations because it lies below the threshold of direct observation and measurement. It is, however, a major problem when investigating reality at the quantum scale. More than a century after Planck's discovery that physical reality is quantized, science still has not fully realized the implications of his discovery.

Mainstream scientists remain unaware of the need to replace or refine the calculus they are using to investigate the quantum world. And while all the data identifying the multitude of particles of the "particle zoo" has been very useful, it has also produced a confused picture of reality, which is a definite sign that something is wrong. Changes must be made in the way we are investigating quantum phenomena. We've solved this problem on the quantum scale by replacing Newtonian Calculus with the Calculus of Dimensional Distinctions (CoDD) and a truly quantized basic unit. But how does this affect studies of the cosmos?

Dark Matter and Dark Energy

In the current mainstream paradigm, the existence of 'dark' components of the structure of the physical universe that cannot be seen because they do not emit or absorb light or other electromagnetic radiation, were inferred from the motions of distant galaxies. Based on Planck Probe data, the Planck Probe mission team, applying the standard model of cosmology, found the proportion of dark matter and dark energy to the substance of the whole universe, to be 95.1%. This figure has been well substantiated by several studies. The calculation is complex and involves assumptions about the ratios of regular mass and energy to plasma and ionized gases in the far reaches of the cosmos, but, effectively, 'dark matter' and 'dark energy' are believed to account for the greater part of the matter and energy in the entire universe.

Given the apparent significance of the role of gimmel in the formation and stability of the building blocks of the physical universe, two important questions arise: 1.) Why haven't physicists found gimmel before this? And 2.) What, exactly, is gimmel?

The answer to the first question is: *They have!* They just haven't realized what it is. How could this be? Why would they not recognize something non-physical? To understand how physicists could discover something this important and not know what they have discovered, we must take a short detour into the history of physics, astronomy and astrophysics.

Physics is defined as the study of matter and energy interacting in space-time. This definition implies that everything about the physical universe can be described and explained with only four measurable variables: mass, energy, space, and time. All other measurable variables are combinations of two or more of these basic variables. When a spiral galaxy rotates in a way that implies there is much more inertial mass than the sum of all the mass that can be observed using telescopes and physical analysis, what is a physicist going to call it? "Missing mass" or "dark matter", of course. As one physicist put it: "We don't know what it is, but we know it has to be matter of some kind." Because of the unspoken assumption that there is nothing but matter and energy interacting in space-time, it simply does not occur to a physicist that there could be something non-physical affecting the angular momentum of a rotating object.

Originally known as "missing mass," dark matter's existence was first inferred by Swiss-American astronomer **Fritz Zwicky**, who discovered in 1933 that the mass of all the stars in the Coma cluster of galaxies provided only about 1 percent of the mass needed to keep the galaxies from escaping the cluster's gravitational pull and flying apart.

In 1965, Vera Rubin, the first woman to break into the old-boys' club of American astronomy, began to work with Kent Ford of the Carnegie Institution in Washington DC. Ford had designed and constructed an image tube spectrograph, a state-of-the-art instrument allowing telescopes to observe objects that were many times fainter than those that had previously been studied. When they aimed the instrument at the Andromeda Galaxy, they saw that the way it was rotating violated Newton's Laws of Motion. While the explanation didn't become clear to them until two years later, the spectrograph printouts represented the first direct evidence of what became known as "dark matter".

Rubin realized that if this "dark matter" was spread throughout the galaxy, then the gravitational force and the orbital speed would be similar throughout. Rubin and Ford had discovered the invisible stuff that influences not only how galaxies move, but how the universe came to be what it is and what it will become. Being trained in mathematical physics, however, it would never occur to them, that dark matter might not be matter at all, but something non-physical.

Let's adopt the working hypothesis that gimmel *is* dark matter along with its equivalent dark energy. This would be consistent with Rubin and Ford's findings, because gimmel exists in every atom in the universe, and in the greatest proportion in hydrogen atoms that make up about 70% of the observable universe. Thus, it would be spread throughout every galaxy. In the TDVP model, non-physical gimmel replaces two very odd theoretical particles of the standard model: gluons at the quantum scale, and the Higgs boson field at the cosmological scale. Both are assumed to be necessary to explain mass. Gluons are assumed to exist to explain the extra mass obtained when quarks combine to form protons and neutrons, and the Higgs boson is assumed to explain mass in general. Both have little or no mass themselves, but somehow are assumed to impart mass to all of the other particles forming the physical universe. The exact mechanism by which a massless particle imparts mass to other particles is unknown.

On the other hand, the existence of gimmel, not measurable as mass or energy, and therefore non-physical, affecting the mass of spinning objects by occupying up to 86% of the volume of the compound spinning objects we call atoms, is a much more elegant explanation.

Based on the discovery of gimmel, a form of the substance of reality that cannot be measured as either mass or energy, we have a very different explanation. It appears that the terms Dark Matter and Dark Energy are *misnomers* arising from the unwarranted theoretical position of looking at reality from a narrow materialistic point of view. We conclude that the 'dark' component of the universe is in fact, gimmel. To perform the appropriate calculation, the data must be converted into volumetric equivalents because the 95.1% proportion is derived from linear mass/energy units. This conversion of the Planck Probe data to volumetric equivalence, yields the volumetric proportion of dark matter and dark energy in the universe of 86.01%.

When the proportion of gimmel to total mass and energy from our TRUE analysis of the elements, using the sum of the percentages of the most abundant elements calculated volumetrically, and using the most accurate figures available, the ratio of gimmel to total mass and energy in TRUE quantum units for the elements and the Planck Probe are almost identical. The

TRUE volumetric percentage is calculated to be 86.09% compared to 86.01%, derived from the Planck Probe data.

We originally proposed that the hypothesis that the dark substance is gimmel, should be considered as likely true, if it were within 2% of the Planck Probe figure, a reasonably stringent requirement. We found only an 0.08% (or 8 in 10,000) difference from the Planck probe figures. This difference is most likely an artifact of measurement sampling error, and the literature on the probe supports this.

Modeling the Cosmos

What does the cube have to do with all of this? It is a 3x3x3 symmetric macroscale object with nine orthogonal planes of rotation. Mathematically, reality is unstable and incomplete without mass, energy and gimmel spinning in nine dimensions. The planes of rotation of the cube can be used to represent the nine dimensions that make up the dimensionometric structure of reality. This structure is analogous at any scale. Symmetric cubes of various sizes can be used to model the quantum world of particle physics, and the macro-scale world the everyday scale of conscious life, and as the application to dark matter illustrates, this can be expanded to apply on the cosmological scale of planetary systems, galaxies and the entire physical universe. The key is to think of the sub-cube as the basic quantum unit of measurement of the building blocks of which all things are made.

As long as the material aspects of reality dominate the thinking of scientists, there will be unresolvable paradoxes in the standard model. The empirical evidence is telling us that the physical universe is only a relatively small part of reality, about 14% volumetrically. There is a much greater reality available to human experience, but scientists trained in the current physicalist paradigm, and most people in modern society, are focused on the material aspects of reality and remain blind to the more important holistic reality, much as if they were asleep. My hope is that we can awaken a few.

PART V: THE ULTIMATE SOLUTION

A Comprehensive Approach to Problem Solving

Solving the cube is not a one-time thing for real cube enthusiasts because there are so many different possible configurations, something new can arise and present a learning opportunity every time one takes a cube in hand and starts rotating the nine-cube tiers. By using the cube to simulate real systems, you can begin to see the ordinary circumstances of life in a new light and learn how to live a more effective and fulfilling life in the process. In this part of the book, I want to share some of what I have learned, in the hope that it may be helpful to someone who is starting out on a journey of discovery using the cube.

As one practices, one become more and more proficient at solving the cube. As I said, it is like learning to play a musical instrument. At first, you memorize the algorithms, practicing them over and over until muscle memory is established to the point where it takes over from rote memory, and you can execute the algorithms with your eyes closed, without thinking about the individual rotations. Like touch typing, or finger work on the piano or guitar, after each series of rotations (algorithms) you automatically know where to go with the next series.

The key is to know exactly what each algorithm will accomplish in terms of moving individual cube faces, and how they may change the configuration at hand to transform the cube.

With practice, you also begin to recognize patterns as they form and anticipate where to go and what algorithm to apply. You become much like the accomplished musician who knows exactly which run or bridge to use to get to the next phrase of melody already anticipated in the musician's mind.

Focused practice of the algorithms is a form of alignment with the algorithms and the geometry of the cube. Once you become consciously aligned, you are set to begin to experience intuitive leaps into new territories, territories analogous to dimensions beyond the 3-D of space and 1-D of time that we are accustomed to.

When the Cube Comes to Life

At some point, a miracle happens: the cube will appear to come to life. As you automatically apply algorithms to a scrambled cube, suddenly the cube becomes a conscious entity, clearly and agreeably helping you to move it toward its perfect alignment. You'll begin to notice that when you finish one application of an algorithm, the cube will be set up for the next. Before you think: "He has definitely gone off the deep end!" Consider this: The discovery of gimmel in every atom of physical reality means that everything possesses some level of, and connection with consciousness. At some point, through focus and alignment, you begin to tune in to the consciousness that exists at the quantum level of all structure. Accept it and go with it. You will find that following the clues presented by the cube will always save you a lot of work in the long run. In general, the cube knows its structure better than you do.

After you have completed a specific series of rotations, an algorithm designed to achieve a certain end, like placing an edge piece in its proper position, don't start turning the cube randomly looking for patterns, that will waste your time and energy. If you have executed the algorithm properly, the cube will already be set up for the next step toward the goal of complete alignment, and the clue leading to the next algorithmic application will be right in front of you! If you miss the clue, you may still eventually solve the cube, but the path by which you achieve it will not be the optimum one. The cube knows the way better than you do; its very being is geometric structure. The same thing is true in life. If you have embarked on a path designed to take you to a higher level of enlightenment, and have completed one step properly, the next step will automatically present itself. The universe knows its structure far better than you do.

Conscious Dimensional Mathematics and Mysticism

"Mysticism" is a term applied to pretty much any system of thought not understood by those in the mainstream institutions of human thought. Believers in the organized institutions of human society, religious, political or scientific, look upon expressions of conscious awareness of the mathematical structure of Sacred Dimensional Reality as fantasy or even heresy. To one on the path to cosmic consciousness enlightenment, any social institution, whether religious, political or academic, is at best, a temporary aid, like a bridge across the chasm of chaos to a comprehensible worldview on the other side. At worst, institutions become authoritarian anchors, tying your body, mind and soul down to mediocracy and stagnation.

Once a social organization or institution has served its purpose, recognize it for what it is, be thankful for it, and move on. There is no need to praise or malign it, just move on. In your unique experience, the bridge may have been a blessing or a curse, but, after crossing the river, you should not allow yourself to be attached to the bridge. Trying to drag a bridge down the road after crossing a river is a form of insanity. It is like continuing to apply an algorithm for the second layer of the cube after the second layer is perfected. It is obviously counter-productive.

We have seen how alignment and intuition are key factors in solving the Rubik's cube, and we've seen how the geometry of the 3x3x3 cube is analogous to the dimensionometric structure of various aspects of the universe, making it useful as an effective model of elementary particles, atoms and elements, as well as human development and dark matter and dark energy. Now, we will try to see how lessons learned in the applications of the cube to simulations of quantum, macro-scale and cosmic reality, including analogies to spiritual experience, can be generalized to apply to the human condition. First, let's look at the deeper meaning and importance of alignment in the context of human consciousness.

Alignment and Intuition

There is a reason that Native American medicine men align themselves with the cardinal directions and build this alignment into their places of worship. Like practitioners of Shamanism the world over, they recognize the fact that the natural physical world is an extension of the spiritual world. In scientific terms, the subtle forces of electricity and magnetism that govern the natural world, involve orthogonal forces, that is, forces that act at right angles to each other, just like the six cardinal directions: North, East, South, West, Up, and Down.

In the wave propagation of electromagnetic radiation, e.g., including visible light, the force of electrical charge and the force of the magnetic field automatically project at 90-degree angles to each other, and the direction of radiation is orthogonal to both of them. This orthogonality is an extension of logical structure of the extra dimensions of time and consciousness, carried into the world of observation and measurement by the conveyance equations discussed in **Part II**.

The nine-dimensional domain of finite reality conforms to the triadic mathematical logic of the conscious substrate because it is the concrete manifestation of that logic in 4-D spacetime, and thus the forces of the natural world conform to the magnetic orientation, spin and motion of the planet. Orienting the cube to the center pieces, which are always orthogonal to each other, is similar to orienting the consciousness of the meditating shaman to the cardinal directions on the face of the Earth.

Alignment, with concentration and focus, creates heightened awareness and eventually an expanded state of consciousness in which *intuition* begins to operate. Intuition involves an expansion of consciousness beyond the 3 dimensions of space and one event in time that we are accustomed to, into a domain where the second and third dimensions of time and consciousness become at least partially revealed. Alignment and intuition allow glimpses of the innate consciousness in all things. The same kind of alignment and intuition developed in learning to solve the cube is also very useful in everyday life.

Synchronicity

The more *aligned* one becomes, the more likely one is to experience amazing synchronicities. Why? Because the world is constructed of patterns and cycles. When your physical, mental and spiritual consciousness is aligned with the geometrical structure, logical and spiritual laws of the universe, you will experience patterns and cycles as a natural part of your life. As your individualized consciousness expands to become more and more aligned and congruent with the primary conscious substrate of reality, you can begin to see that everything happens in accordance with the laws of the universe. Then, the so-called "Secret Law of Attraction" will begin to work for you.

Look for unusual events that are clues naturally arising in the course of your life on the path to enlightenment. They may be signs indicating a future event that is days, weeks or months away, or they may be immediate. They are like the clues the cube presents when you are on the right track to restore it to its perfect chromatic coherency. They are clues to what is coming next. Accept them and go with them, and your path will be smooth, resist them and your path will become more complicated, frustrating, and maybe even dangerous and painful.

Personal Experiences

In 1955, when I graduated high school, the Korean war was going on, as was the cold war with Russia, and the US armed forces were obtaining reinforcements through the draft. Many of my high-school classmates volunteered immediately after graduation to avoid being drafted. By volunteering, they hoped to choose the branch of the military in which they wished to serve. From 1951 until 1955, I had experienced a number of very unusual events that suggested that I would never be in the armed forces and that I would become a scientist, so I did not volunteer. I had a couple of scholarships, so I entered college, majoring in physics and mathematics. But in 1958, my roommate and I, disillusioned with academia, decided to drop out, even though we were both 'A' students, and started a hitch-hiking journey across country on a spiritual quest. Our parents, nearly had heart attacks.

My roommate entered a monastic order, and I returned to Missouri, where my college girlfriend and I got married in 1960. After a few months, we moved to Los Angeles, where I was initiated into an ancient form of consciousness expansion known as Kriya Yoga. (More about the spiritual journey later.) In 1962, we returned to Missouri, so I could finish my bachelor's degree. The war in Viet Nam had started, and the cold war with Russia had heated up with the Bay of Pigs Incident in 1961. I enrolled in College, but before classes started, I received a letter from the Department of Defense that began with "Greetings! Report to Jefferson Barracks in St. Louis, Missouri for examination and induction into the United States Army."

I informed the draft board that I had enrolled in college and paid a semester's tuition *before* I received the draft notice. They said that made no difference. I would have to report for examination on the date indicated in the letter. The draft was mandatory in those days, and if you were physically fit, you had no choice. There was a moment when I wondered how this would impact my life, but when I went into deep meditation and asked about it, I was shown a chain of synchronistic events that told me that I would not be a soldier. There was nothing for me to do but go with the flow and let the universe do its thing.

When I reported to the induction center, I received a page of instructions and was directed to go to the back of a line. I was in good spirits, feeling great. Standing in line with the other inductees, I chatted with the young man in front of me. I could tell that he was nervous and not too happy about being there. I couldn't explain to him why I wasn't worried. I had had glimpses of my future, and military service wasn't part of it; and I had learned to trust my internal guidance, regardless of the outward circumstances.

I passed the physical exam with flying colors and was directed to get dressed and stand in another line to take an aptitude test. I was good at taking tests, so that didn't bother me. It was a general aptitude, math and word recognition test somewhat like the standardized IQ test I had taken in high school. We were told to sit on chairs spaced well apart along a long narrow table facing a blank wall. The test folders were passed out, and we were told to wait until the proctor said GO, open the test and start. I finished it quickly. I think I was the first to hand it in and leave the room.

I was asked to wait in the room just outside the examination room, so I sat down and read a brochure with Uncle Sam pointing his finger at me on the front, over the words: "Uncle Sam wants YOU!". After most of the recruits had emerged, a man in uniform came out and called my name. I raised my hand and he motioned for me to come back into the exam room. When we were inside, he said:

"You're going to have to take the test again!"

When I asked why, he didn't answer. He just motioned for me to sit down and placed a fresh test folder in front of me. He then pulled out a stop-watch, said GO, and stood over me until I had finished the test. The test was essentially the same test I had taken before, just rearranged. Of course, it took less time this time than the first time. He took the test and disappeared into an office nearby. A short time later he returned and said:

"You can go now."

I asked him again why I had to take it over,

"Was the first test sheet messed up or something?"

He glanced over at the office door, drew closer, put his hand up beside his mouth, and whispered:

"I'm not supposed to tell you this, but – your score was *too high*. The proctor thought you must have cheated!"

Several days later, I received a letter from the President of my Draft Board saying that I could enroll in college. The letter said that the county where I attended high school had ample volunteers at that time, and that, as long as I maintained grades of B or above, I could disregard the induction notice.

I had enrolled in a heavy load of classes, upper-level classes in advanced math, physics, German and logic, and I left my wife behind in Southern Missouri for the first semester and rented a small upstairs room at the edge of campus and applied myself diligently. At the end of the semester, I all A's and B's.

After graduating, with a degree in mathematics, I taught high-school math in Newberg, Missouri for two years. I enjoyed teaching very much, and the first year, my students won first place in the Regional Math Competition, even though we were up against several much larger schools. But the pay earned by secondary school teachers in those days was borderline poverty, and public education was becoming politicized by liberal professors in the education departments of our universities and the teachers' union. There was no longer a path for exceptional students. the three-track system had been abandoned, and the brightest students were mixed in classes with the average and the slowest learners. This, in my opinion, was the beginning of the inevitable dumbing down of public education.

When the first five-week evaluation period rolled around, and I posted the failing grades that two or three of my students had earned, the Principal informed me that I could not fail students. I had to pass all students, whether they learned the material or not. Fighting with the dumbing-down of the American educational system for minimum pay was not the way I wanted spend my time and energy; so, by the second summer break, I had had enough. I decided to look for gainful employment elsewhere. Eventually, I found a job with the US Geological Survey as a hydraulic technician in the Water Resources Division. It was this job that led to my work with the Department of Interior Systems Analysis Group, a water resources modeling study in Puerto Rico, and ultimately my PhD in environmental engineering.

The universe, or God, if you prefer, knows what it, He/She, is doing. But as limited individualized conscious entities, we cannot see the whole picture, and the way is not always clear. The path to enlightenment and the final goal of Cosmic Consciousness may not be obvious, and we are constantly confronted with choices. Like the choices of many possible rotations with the cube, many choices lead to trouble, some even to death and destruction, and like the cube, only a few are correct choices, and the optimum path to spiritual growth and ultimate freedom is difficult to find.

It helps to realize that the river always flows to the sea. Your body-mind-soul is like a leaf floating on the water, tossed and swirled, sometimes submerged, but it will eventually reach the ocean, however battered and broken, because that is where the river is going. There will be eddies and sloughs along the way, where you will languish, until a flood comes, unless you make a mighty effort to get back into the flow. The trick is to align yourself, body, mind and soul, with the flow. Once in the flow, you must go with the flow. To fight against it is folly.

Sometimes the need to go with the flow is *immediate*. For example: One afternoon in the El Junque Rain Forest of Puerto Rico, I was carrying my four-year-old adopted son Dale, across a stretch of flat rock to access a path leading to a place where we could jump or dive into the sparkling clear water of a pool below a waterfall on a mountain stream called *Sangre de Los Santos*. The footing became unexpectedly wet and slippery and suddenly, I felt myself sliding toward the edge of the rock where there was a straight drop of 40 or 50 feet to a field of jagged rocks and boulders waiting below.

If I had tensed and resisted falling, I wouldn't be here to tell you about it today. But I had been meditating on a daily basis for several years (more about this later), and I knew that everything happens for a purpose. If this was to be the end of this life, I would throw my son away from the cliff and fall. But first, I would just relax, let the universe do its thing, and go with the flow. I relaxed completely and fell. A peaceful calm came over me as I allowed myself to fall onto the slippery rock. I landed on my side and turned onto my back. This increased the area of skin sliding on the rock enough so that the friction brought my body to a halt at the very edge of the precipice. This would not be the day I would die. Going with the flow had saved my life.

Belief Versus Knowledge

Many people approach life haphazardly, as if there were no universal laws, no meaning, no purpose, making choices without thinking much about goals or consequences. With this approach to life, the road to final enlightenment is probably going to be long and uncertain. So, how are we to proceed? Institutionalized religions offer ways that require little or no thinking. Just accept the doctrine and follow the rules, which, they may claim, were laid down hundreds or thousands of years ago by a fully enlightened being. But, the teachings of enlightened beings are often distorted by lesser beings who fail to understand, or even worse, who choose to use religious doctrine to gain power and control over people. In such cases, religion becomes a form of political machination and the true spiritual nature of religious practice is lost.

During the past 100 years or so, simple physical science has exposed the teachings of many religious beliefs to be false, and because of this, some people have concluded that all religion must be false. Not realizing that mainstream science has ignored the fact that there is an unspoken metaphysical basis underlying any system of thought, they have simply traded belief in religion for belief in materialism. In effect, science has become the new religion. Materialism, not science, has become the new religion for many, and mainstream materialistic scientists are their priesthood, but materialism is not science. In fact, the assumption of materialism is not a valid scientific hypothesis.

Materialism is based on the belief that reality is nothing more than matter and energy interacting and evolving in space and time, that we are nothing more than an accidental combination of debris flying away from a big bang explosion. But, we noted in **Part II** that this does not meet the criteria of a scientific hypothesis. Materialism is not science. By rejecting religion entirely, materialists have virtually thrown the proverbial baby out with the bath water. They have thrown Jesus, Buddha, and all enlightened beings out with the organized institutions that have distorted their teachings!

So, if the dogmas of religious institutions are false, and mainstream science is unscientific, where do we turn for answers? There are an astounding number of people in the world today who style themselves as "life coaches" or experts of some sort, who believe it is their mission to enlighten you with their clever advice, for which they believe you will pay a lot of money. Avoid most of these people; especially if you can see that their personal lives do not reflect alignment and harmony with natural law.

I'm reminded of a preacher in the small town where I grew up, who after being seen drunk on Saturday night, would say on Sunday morning: "Do not as I do, but do as I tell you to do"! What to do? Everyone appears to be crazy, except you and me, and I'm not so sure about you! Where do we go from here?

You've probably heard someone say: "Everyone must have something to believe in!" And the person giving this advice is probably sincere, but be very careful, there is a vast difference between 'belief in' and 'knowledge of' something. What most people mean by *belief*, is a statement of what they would *like* to be true. Ask the giver of this advice: "What do *you* believe?" You will probably get a recital of a church doctrine, or a philosophy of some kind. The next question to ask is: "Do you know with absolute certainty that this is true?" If they are honest, they will probably say: "No, but I *believe* it to be true".

That means that their "truth" is a hypothesis, not a statement of direct knowledge. Until you have proved a hypothesis for yourself, with mathematical logic and direct personal experience, it is still just a hypothesis. You should strive to replace "I believe" with "I Know" whenever you can, because unless you can replace blind belief with certain knowledge, you are relying on someone else's knowledge, or worse, on their unfounded belief of something that is untrue, and in that case, your belief is nothing more than wishful thinking.

Finding the Path

Are there people who *do know*, i.e., people with direct knowledge and proof? If not, we're all in deep trouble! Yes, there definitely are truly enlightened beings who know what they know from direct experience and indisputable proof. But they are few and far between, and to meet one is a rare privilege. How do I know this? I have met a few. I will share some of these meetings, but first, I want to address a very serious problem that is a barrier to real knowledge for many people today: the fallacy of the assumption of absolute relativism.

This belief is rampant in so-called "New-Age" thought. It is not uncommon to hear: "Oh, yes, Einstein proved that everything is relative"! This promotes the idea that there is no absolute truth, and therefore, we are free to believe whatever we want to believe. Of course, we *can* believe whatever we want, but we do so at our own peril. Belief in a false concept is the door to endless wandering in self-indulgent delusion, and eventually leads to trouble and disillusionment. Reality is governed by very specific invariant laws. Actions counter to those laws create serious suffering.

If Albert Einstein could hear someone saying that he proved that "everything is relative", he would be turning over in his grave! He was a firm believer in an ordered universe that conforms to definite mathematical logic, and a set of precise rules known as the laws of physics. He proved that the measurement of space, time, matter, and energy, for different observers, depends upon the relative motion of the observers, and he provided equations that determine exactly what the differences are, given the velocity of their relative motion. Those equations are known as transformation equations because they transform the measurements in one reference frame to those of another, when the reference frames are in motion relative to each other. All of this is determined based on physical laws and mathematical principles. In no way did Einstein prove that *everything* is relative.

Fortunately, there are a number of things that are absolute in the universe of our direct experience, including at least five physical constants, including the speed of light in space. No science or technology would be possible in a world where everything is relative. Orderly patterns in such a world would quickly decay, break down and self-destruct, leaving a world of complete chaos. This is not a world anyone should wish to live in.

Be advised that there is a consistent reality. Scientific theories based on various hypotheses will contain some truths, but we've seen a number of scientific theories come and go in just the past 100 years. Scientific theories may be true, false, or partly true, but hypotheses proved using mathematical theorems are changeless truths. For example, the length of the longest side (called the hypotenuse) of a right triangle consisting of three straight lines, will always be equal to the square root of sum of the squares of the other two sides (the Pythagorean Theorem). Nothing will change that, because it is not relative, it is mathematically invariant.

Fermat's Last Theorem for cubed integers, as used in **Part II** of this book, is as true now, as it was when it was first proved by Pierre de Fermat in 1637, yours truly in 1965 and Andrew Wiles in 1994. Similarly, the proofs of Euclid's Elements, proof that there are only five regular solids, and proof of Gödel's incompleteness theorems, all are as absolutely true now, as when they were first proved, and they will be true forever. Mathematical truth is not something invented by human beings, it is, and always has been, there to be discovered. An alien life form would discover the same mathematical truths, only the notation, dependent on variances in modes of expression, might be different.

Enlightened Beings

It is the existence of absolute truths that makes it possible for there to be a few real human beings who have attained a level of enlightenment and realization beyond the ordinary. Because our physical senses are limited, there are a finite number of absolute truths in the world of our direct experience. So, a person who has direct knowledge and proof of the knowable truths in this universe can be called a fully enlightened being. Such a being would be the personification of Cosmic Consciousness, a term coined for the next level of the evolution of the human mind, by Dr. R. M. Bucke, a Canadian psychiatrist, in the 1890s. Have you ever met such a being? Have I? Would we recognize one if we did? Most people are much too busy with what they believe, and mistakenly think they know, to be able to recognize such a person.

Remarkable People

My first meeting with someone who may have had abilities beyond those of the average person, occurred when I was only a few days old. My father had an aunt who never married. In those days, a lady who remained unmarried past the age of 30, was referred to as an "old-maid". My dad's old-maid aunt came to help my mother with cooking and cleaning when I was born. She took one look at me and exclaimed:

"This child will be famous!"

I was born with curls of black hair, and a lock of it hung down on my forehead. It was that "forelock" that my great aunt said foretold a life of fame. I also began speaking in four- or five-word sentences within a few months. My maternal grandmother told my mother that she feared I would not live to be grown. She said: "He's too smart to live!"

Apparently, and fortunately for me, my grandmother was not psychic. I enjoyed a largely untroubled childhood in Arcadia Valley Missouri, a typical American small-town. I was an only child, but I had plenty of friends and cousins.

A Remarkable Childhood Friend: Robert Dettmer

My best friend as a child, was a classmate named Robert Dettmer. By the age of seven or eight, we were writing poetry, learning German, inventing secret codes, and using tomato juice as invisible ink. A message written in juice would dry and remain invisible until heated from beneath the paper with a match. The heat would turn the dried tomato juice brown. Once you had read the secret message, you could burn it. Junior spys!

We built model airplanes, crystal radio sets, and ordered books on things as disparate as farming, hypnosis and body-building, from magazine ads. Our friendship lasted until my parents moved to Phoenix, Arizona in 1949. After high school, my friend Robert went to medical school at the University of Missouri, and later returned to the Valley to practice medicine.

An Inspirational Speaker: Bob Proctor

I met Bob Proctor in 2012 in Salt Lake City Utah, where he was the keynote speaker at the Young Living World Convention. Bob is a disciple of Napoleon Hill and Earl Nightingale. He has been very successful as an entrepreneur and inspirational speaker for many years. He was featured in the inspirational movie <u>The Secret</u> in 2006. He teaches excellent courses in self-help and motivation and has inspired many people to change their lives for the better and become more successful than they ever thought possible. Meeting Bob Proctor qualifies as one of my meetings with remarkable people.

Is Bob an enlightened being? I'm sure he would say no, but he certainly has reached a level of understanding of how things work, that goes beyond that attained by most of us, and his success is based on the hypothesis that there are in fact universal laws, including the Law of Attraction, which was highlighted in <u>The Secret</u>. He teaches that by following universal laws, one will obtain specific predictable results. As you might have gathered, I agree with him.

The theme of Bob's keynote address at the Young Living Convention was the importance of determining your purpose in life. Jacqui and I were in Salt Lake City for the convention, and when I saw Mr. Proctor in the Marriott Hotel Restaurant before the convention, I walked up and introduced myself. I informed him that I already knew my purpose in life, which was, and is, to bring science out of the dark ages of materialism, and I gave him a copy of my book <u>Transcendental Physics</u>. We talked for a few minutes and he asked me to come back stage after his presentation. Backstage, after his inspiring talk, he gave me a copy of his latest book and we vowed to keep in touch.

But, let's go back in time a bit: Long before meeting Bob Proctor, I had already met a few people who were more highly evolved both mentally and spiritually than most of the inhabitants of this planet at this time.

A Very Remarkable Friend: David Stewart

In 1955, I met David Mack Stewart, one of the most remarkable people I know, it was our freshman year in college. We happened to be in rooms across the hall from each other in the freshman men's dormitory. He was a pre-theology philosophy and English major, I was studying physics and mathematics. One would think that we shouldn't have had much in common; but there was a connection much deeper than anything physical or intellectual. We could read each other's thoughts. We would often complete each other's sentences, and occasionally say a complete sentence in unison.

More than once, while sitting in a physics or math classroom, listening to the lecturer, working on problems, when suddenly, I would get a mental image of Dave. I'd look up, and he would be passing by a window of the Science Building, on his way between the dorm and the Liberal Arts buildings at the other end of campus.

In late night discussions, which were common in the freshman dorm, we found that we had very similar understandings about philosophical subjects, and a lot of ideas that neither of us had ever been able to share with anyone before. Dave's brother Ted said we were twins who just happened to be born to different mothers. David and I became lifelong friends, and embarked on a spiritual quest together, as mentioned in the section on personal experiences of synchronicity above.

A Remarkable Scientist and Mathematician: Norbert Wiener

When I was a sophomore in college, I had the privilege and honor of meeting a remarkable scientist: Dr. Norbert Wiener, PhD. He was a child prodigy who grew up in Columbia, Missouri, where his father was a professor. He earned his BA in mathematics at Tufts University at the age of 14, and had attained world-wide recognition as the father of cybernetics by the time I met him in 1957.

Because I was a member of Kappa Mu Epsilon, the National Mathematics Honor Society, I was privileged to be one of a small group of students for whom he explained the mathematics of brain waves in a meeting at the University of Missouri. I was able to talk with him and ask questions. His answers were food for thought. His mathematical description of brain waves suggested to me that there could be real non-physical aspects of human consciousness.

A Brief Encounter with a Remarkable and Unusual Man: Dr. Mahija

While I was in college, a friend took me to meet an interesting man in Kansas City, Missouri, a man who he believed to be an exceptional genius, a doctor who had cured many people diagnosed with terminal diseases. The man was a black man known to my friend only as Dr. Mahija, and he was rumored to be well over 100 years old. I took advantage of a long weekend to visit my friend in Kansas City, with the possibility of meeting Dr. Mahija.

We waited outside his office for a while and were able to have a brief interview with him between appointments with patients. He ushered us into his laboratory, which was adjacent to his office. His lab resembled a chemistry lab, but was filled with strange equipment I had never seen before. Sitting on a table in the middle of the laboratory, bathed in sunlight from a skylight, there were jars of colored liquids in which spheres of silvery substances of different colors were suspended. I asked what they were. He responded that he was distilling Sodium, Potassium, Mercury, Platinum, Silver, and Gold.

"From what?" I asked. His reply shocked me:
"From sunlight and the atmosphere".

Before I could ask another question, his receptionist came into the lab to inform Dr. Mahija that his next patient had arrived. I said that I would like to return sometime and learn from him. He paused at the door, turned and said: "You'd be welcome, but most people know too much to learn anything from me!"

Was Dr, Mahija a remarkable person? Probably so. I regret not having found the time to return and learn from him. Who knows what I might have learned?

Embarking on the Spiritual Quest

Late one night in 1958, in our dorm room, David Stewart and I made a big decision. It was not an easy one. It was a decision that would radically alter the courses of our lives. In our junior year, even though both of us were honor students, we decided to drop out of school and seek the truth elsewhere. We had discovered that there was much more to reality than was being taught in the classrooms of the colleges and universities of the world. Conventional education was not enough for us.

Dave was a philosophy major, planning to become a Methodist minister. I was a physics major, planning to become a theoretical physicist. Our backgrounds were similar in that we were both from small Southern Missouri Ozark towns, we were both honor students in high school, and we had both had remarkable personal experiences of spiritual and psychic phenomena. Dave's family were Methodists, but my family background was mixed with regard to religion. One side, my father's side, were church-goers; my grandmother was an organist for a local church, but my mother's family were not church goers, my maternal grandparents were decidedly agnostic. But my mother did her motherly duty, taking me to Sunday school at the local Baptist Church.

At about the age of 12 or 13, I walked down the aisle of a Landmark Baptist Church and offered my life to Christ. But I did not join the church. I saw most of the members as hypocrites. For them, spirituality was limited to one day per week: Sunday. The rest of the time, they seemed to forget about God entirely. Church was just a social organization for them, a place where they could meet and pretend to be spiritual. Even at that age, I felt that if spirituality was real, then it was something that should be part of your life every moment of every day.

When I graduated high school, I was awarded scholarships by several colleges and universities, but I chose Central Methodist College, not because of religion, or because it was a church-related school, but because of its excellent accreditation and faculty, its small classes and, most importantly, because when I visited the campus, I felt at home.

Experiments in Parapsychology

As roommates in our second and third years, Dave and I had become interested in the budding science of parapsychology, headed up at that time by Dr. J.B. Rhine at Duke University. We designed some experiments to investigate a type of psi phenomena that we knew from personal experience to be real. We were encouraged and supported by two well-respected professors, one in the English Department, and one in the Physics Department.

One of our experiments was very successful, demonstrating the ability to obtain specific knowledge we could not access by normal means. It yielded consistent repeatable results indicating either clairvoyance, telepathy, or contact with spiritual entities. We documented the results and sent them to Dr. Rhine. Dr. Rhine encouraged us to continue our research, but wisely advised us to obtain degrees that would enable us to find work, because there was no real demand for parapsychologists. We were ignoring his wise council regarding completing our degrees, at least for the present.

We left school, back-packed into the wilderness of the South-Central Missouri Ozarks and camped in the mouth of a large cave that I knew about in a remote area of the Current River Basin. We took only the bare-essentials with us into the wilderness: sleeping bags, hiking boots, a change of clothes, an axe, a 22-calibre rifle, fishing tackle, pocket knives, salt, lard, flour, and several good books. There was a crystal-clear spring in the cave, and a spring-fed stream ran a few feet below the mouth of the cave.

The Mouth of the Cave: Our Campsite

Big Creek

We planned to stay there for at least three weeks, relaxing, reconnecting with nature, and deciding how to proceed. We spent the days hunting, fishing and swimming, the afternoons reading, and the evenings sitting around the campfire in the mouth of the cave, discussing philosophy and the meaning of life.

We used methods we had developed in our psi experiments to psychically locate things like unusual food sources, American Indian artifacts, firewood, and old tires to burn in our campfire to provide more light for reading in the twilight and even into the night. One night, during the third week of our stay, I had a lucid dream. In the dream, I walked into the cave. The floor sloped up to a room that curved to the left. At the far end of the room was a steep jumble of rocks that appeared to have fallen from the ceiling.

I was surprised to see a light glowing from an opening at the top of the jumble. As I approached it, the jumble of rocks turned into a flight of stairs, and I could see that the light was coming from a door at the top. I climbed the stairs and stepped into the room. A young man with dark skin and black hair, probably only a few years my senior, sat cross-legged on a slightly raised portion of the floor. He had a sparse, well-trimmed beard. When I looked into his eyes, he smiled, - and the dream ended.

The next day, we decided we would return to our parents' homes for a few days and prepare for a hitch-hiking journey across the country in search of the spiritual connections we knew were out there. We spent one more night in the cave, and that night it rained. The next morning, we were surprised to see that the mummering brook that ran in front of the cave had become a raging torrent of muddy water. The log we had placed across the branch to reach the cave was gone. Muddy water, fifteen feet deep and perhaps a hundred feet wide confronted us. Logs and large up-rooted trees were floating by. There was no way we could cross the creek to reach the trail leading back to civilization. It was as if nature were conspiring to keep us there!

That afternoon, my dad, who had come to check on us, hailed us from across the roiling water. Yelling back and forth over the roar of the flood, we decided we would climb the bluff and hike several miles to the small town of Eunice, on our side of the creek, while my dad would go back out and drive around via the Route-17 bridge, several miles up-stream, to Eunice to pick us up. After a few days' rest, we decided that we would hitch-hike to California to meet famous people and look for the enlightenment we didn't think was available in school.

As it turned out, Dave was more committed to this part of the quest than I was. He sold or gave away everything he owned, packed his backpack with extra clothes and snacks and donned his Eagle Scout uniform (In those days, drivers were more likely to stop for a Scout) and we met in Jefferson City to start the next leg of our journey. The first night we slept in our sleeping bags in a city park, the next night in a frat house on a university campus in Kansas where some of Dave's high-school friends were going to school. The third night we slept under an overpass somewhere in Western Kansas. On the fourth day, I decided to turn back. With only fifty cents in my pocket, I hitchhiked back to Central Missouri, where my girl-friend, an art major, was still attending classes. Dave went on to California.

The Spiritual Quest Continued

In 1959, I moved to Kansas City, Missouri, where I got a job selling vacuum cleaners. I had continued dating Bobbi, and I proposed that we get married. She turned me down the first time, but I persisted, and after a couple more proposals, she accepted, and when she graduated from college, we were married in St. Louis, where her parents lived, on June 20th, 1960, and we rented a house in Kansas City, where I was doing well as a salesman.

About a month later, I received a package from Dave, who was still in California. It was a book entitled "The Holy Science" by Swami Sri Yukteswar Giri, an Indian sage. It was thoroughly marked up, with highlighted passages and Dave's handwritten comments in the margins. After reading the book, and a small booklet called Highway to the Infinite, also sent by my ex-roommate, I told my new wife that I wanted to move to Southern California to study the teachings of an Indian Saint. To my surprise, she agreed. I think she agreed because it sounded like an adventure. We packed up everything we owned and headed for California.

Arriving in Los Angeles in July 1960, we drove straight to the Headquarters of Self-Realization Fellowship atop Mt. Washington, just north of downtown LA. As we drove onto the grounds, just inside the gate, our car, a Studebaker Golden Hawk, that had functioned perfectly well all the way from Missouri, died and refused to move any farther. While one of the monks started working on the car, we met with Dave and some other disciples of Paramahansa Yogananda, a disciple of Swami Sri Yukteswar Giri. These people, including Dave, considered Yogananda to be an Avatar, a Sanskrit term which meant a *fully enlightened being*. They also considered Sri Yukteswar and his guru, Lahiri Mahasaya, disciples of a Himalayan Yogi known only as Babaji, to be fully enlightened spiritual masters.

Sri Yukteshwar Giri

Was this some anti-Christian cult? I knew in my heart that it was not. Dave was a dedicated Christian, planning to be a Methodist minister; a Methodist through and through! And there was a large picture of Jesus Christ prominently located on the SRF altar. Membership in Self-Realization Fellowship did not require renouncing your religious affiliation or belief.

Meetings with More Remarkable People

After meetings with a senior monk named Brother Anandamoy, who impressed me greatly with his calm matter-of-fact air and depth of knowledge, I began studying a set of lessons written by Yogananda, and on September 17, 1960, I was initiated into the practice of Kriya Yoga, an advanced form of yoga designed for calming the mind and gaining control of the subtle energies of body and mind. The person performing the initiation was the President of SRF, Sri Daya Mata, affectionately known as *Daya-ma* by SRF members.

During the ceremony, which was truly a life-altering experience for me, the energy centers of my body merged to form a blazing ball of golden light, appearing to be about the size of a baseball, spinning rapidly in mid-air a few inches in front of my eyes. I knew intuitively that concentrating my attention on it would expand my consciousness to encompass an awareness beyond anything I had ever known before. I had found my spiritual path!

After the initiation ceremony, we filed out of the chapel on the ground floor of the SRF Headquarters building and were directed to the stairs to the second floor. As I passed Dave, who was one of the renunciants directing us toward the stairs, I experienced a strong feeling of *déjà vu*. There was something very familiar about the stairs! At the top, I could see a light coming from a room that turned out to be the room where Yogananda had lived until his passing in 1952.

When I stepped into the room, there was a young Yogananda, looking unlike any picture of him I had seen in the SRF literature, sitting cross-legged on a raised area, smiling at me, and I knew at once why I was feeling the strong *déjà vu*. This was the young man with dark skin and black hair I had seen in my dream in the cave in the Ozarks, more than four years before. For a moment, I thought he was alive and three-dimensional, sitting there before me, smiling. I blinked and realized it was just a picture.

We found a cabin to rent in Topanga Canyon, in the Santa Monica Mountains west of LA and started looking for employment. Answering ads in the Los Angeles Times, I took a test, underwent a psychological profile and was hired as an actuarial mathematics technician writing programs for the Univac Computer in downtown LA, and Bobbi found a job working for the telephone company in Santa Monica. We commuted from Topanga, to our jobs in Santa Monica and downtown LA, and I went to Mt. Washington to meditate with the monks in the evening once or twice a week. But, after a few months, we decided to return to Missouri, so I could complete my degree in mathematics and physics.

Continued Education

My interests in mind/matter interaction and psi phenomena, propelled me on a personal search of comparative religions, linguistics, symbolic logic and philosophy. Bobbi was not very interested in these things, but was reasonably tolerant of my involvement in them. My interest in these subjects was such that when I took the graduate record exam (GRE), even though I was completing a degree in mathematics and physics, I did the specific-field part of the exam in philosophy. Based on the results of the GRE, I was accepted into MENSA and a PhD program in symbolic logic at Tulane University. However, due to poor planning and personal circumstances, I never attended Tulane. Instead, I enrolled in a graduate program in theoretical physics in a school close to home, where I focused on mathematics, electronics, and quantum mechanics.

During the next several years, I studied physics and consciousness expansion and taught mathematics to earn a living. As mentioned above, my first teaching job was as a high school math teacher. Although I enjoyed teaching very much, after two years, I realized that teaching paid very poorly compared to professions in engineering and applied mathematics. So, I found a summer job as a seismograph technician with an oil exploration company in Texas, after which I landed a permanent career-path job with the US Geological Survey.

I started with the Water Resources Division of the USGS in Rolla Missouri, and after a few months, was transferred to Iowa City, Iowa for an 18-month hydrologist training program. I was assigned a desk in an office in the Hydraulics Lab overlooking the Iowa River, and another desk in the Geology Building, both on the University of Iowa campus. Most of the first 12 months of training was field work: stream gauging, sampling and aquifer pumping tests, but because of my background in math and physics, I became involved in surveying for indirect calculations of streamflow, the modeling of streamflow, groundwater aquifers, and the interaction between groundwater and surface water.

A Remarkable Encounter with a Psychic Grandmother

At night and on weekends, I meditated and studied consciousness expansion. Also, because I believed that under controlled conditions, certain psychedelic substances might expand human consciousness to include real spiritual experiences, I studied scholarly research papers on psychedelic drug experiments in the university libraries. I studied all the literature I could find on psychedelic drugs like LSD, mescaline and Psilocybin.

I read everything written by Aldous Huxley, David Alpert and Timothy Leary, but I never took psychedelic drugs myself. I considered the risk far too great. The possession of psychedelic drugs was illegal, and little was known about the side effects on mental and physical health. However, I got to know an artist who did take psychedelic drugs and I picked his brain to learn about his experiences with them.

During this time, I was also studying everything I could find on psychics. I knew that most were fakes, but, if *even one* is real, human consciousness has potentials about which science knew next to nothing. From some friends, I heard about a lady in Cedar Rapids, who, according to people who had had readings from her, was remarkable. A geologist I worked with said she told him things that no one could possibly have known about him. Cedar Rapids was a short drive from where we lived, so, one weekend, I made an appointment.

When we arrived at her house, I expected to see a sign like palmistry and tarot readers have outside their places of business in large cities, but this lady's home was a typically modest Iowa farm home, on the edge of Cedar Rapids. And her name was not Madam Zelda, or anything like that. She had the extraordinarily ordinary name of Mrs. Johnson. And apparently it wasn't a pseudonym, because the name stenciled on her mailbox was "M Johnson".

Bobbi and I climbed the steps onto the front porch of the white clapboard house and I knocked on the door. The lady who opened the door looked nothing like a Gipsy fortune teller. She was a typical Iowa grandmother, probably in her late sixties or early seventies, wearing comfortable shoes, a flower-patterned dress and an apron! She invited us into her living room, where a picture of Jesus hung on the wall. There was a cup of tea and some cookies.

Bobbi was not interested in a reading, so after some pleasantries, a few sips of tea and a delicious home-made chocolate-chip cookie, she supplied Bobbi with a magazine to read and ushered me into her "reading room", where she pointed to a couch with a plastic cover, and sat down in an over-stuffed arm-chair facing me. She took my hand briefly and looked at me over dime-store reading glasses, and told me some things that made absolutely no sense to me ... at the time.

"You will soon be having car trouble. I see red flashing lights, something to do with oil, and I know you don't think so, but you'll have to sell your car. And then, I see you surrounded by boxes, boxes of light fixtures stacked up all around you."

She paused and said: "I also see a rushed trip to the hospital, but it's not for you, and it will be OK. Maybe a false-alarm." All of her pronouncements were simple and matter-of-fact, as if she were telling me about the farm report for the corn market.

She told me a few other minor things she "saw" about my personal future, and about world conditions, like the future of the war in Vietnam. I thanked her, and left thinking that this was a waste of time. But I didn't mind giving her the money (I think she charged $20) because I suspected she was on a fixed income.

A few months before this visit, I had purchased a wonderful car, a slightly used 190 SE Mercedes Benz from a retired University Professor. It was canary yellow, with tan, real-leather upholstery and solid wood dashboard with stainless steel toggle switches and trim. The motor purred. You could barely hear it running at a stop light, and there was no vibration, I loved that car, and planned to keep it for a long time. I thought: There's no way I'm going to sell my car!

A couple of weeks after visiting the grandmotherly psychic, we packed the car, late on Friday after work, and headed south to visit Bobbi's family in St. Louis, and my parents in southern Missouri. It was dark before we left Iowa City. Only a few miles out of town, the engine shuddered, died, and the car rolled to a stop. I sat and stared at the red light flashing on the dashboard in the gathering darkness. It was blinding! We had to be towed back to a garage in Iowa City.

Monday morning, when I opened the door to the Special Studies Office in the Hydraulics Lab, I was shocked to see that I had to make my way between shoulder-high stacks of boxes to get to my desk! The boxes were full of six-foot long light fixtures and fluorescent light bulbs! The University was replacing all of the 20-year-old light fixtures on our floor of the building.

A few months later, another of Mrs. Johnson's predictions came true. But it was much more serious than it seemed to be in the reading. Bobbi was in the early stages of a pregnancy, we were expecting our first child. One evening, all of a sudden, Bobbi was in a lot of pain. Rushing her to the hospital, we learned that the pregnancy was ectopic. The fetus had formed within a fallopian tube. After surgery, the doctor informed us that Bobbi would be unable to conceive again. Bobbi was depressed. We had been excited about becoming parents. Now, that would not happen.

More Education, More Meetings with Remarkable People

After the 18 months of field training in Iowa, because of my math and physics training, I was one of a handful of USGS employees selected for intensive training in systems analysis and mathematical optimization techniques at the University of Arizona in Flagstaff, after the completion of which, in 1967, I was transferred to Sacramento California as project engineer on a USGS/US Army Corps of Engineers "Worth of Streamflow Data for Reservoir Design" study in the Corps' Hydrologic Engineering Center.

I contacted SRF to ask if there was a meditation group that I could attend in the Sacramento area, and was informed that the group that had been meeting in Roseville, Northeast of Sacramento for years, was looking for a new place to meet. Would I consider having them meet in my home? I agreed and became the leader of a small meditation group. While in Sacramento, we adopted a five-week-old baby. We named him Martyndall, a combination of two family names: Martz and Tyndall, but pronounced it "Martindale".

About half-way through the research project, the USGS sent me to UCLA for a course in hierarchal modeling. While in

LA, I managed to schedule a meeting with Sri Daya Mata, the President of SRF who had initiated me into Kriya 7 years earlier. I had several questions I wanted to ask, but when I was ushered into the room where she received people, I couldn't think of any of them. After greetings and a few minutes of silent meditation, she said:

"I see you are interested in paranormal phenomena."

I hadn't told Daya ma about my research into altered states of consciousness and psychic phenomena, and I thought I might be in trouble because I knew about Yogananda's teaching about the paranormal, psychic phenomena, self-hypnosis and the like. His position was very clear: they were to be avoided as dangerous and counter-productive to spiritual growth. I swallowed and said:

"Yes, it's true, I am interested in such things. But, ..." I started to say something in defense of my interest, but she held up her hand.

"As you know, Master advised SRF students to avoid involvement with the occult, and that's our general policy. But, in your case, it's different: you are approaching these things from the viewpoint of a scientist; for you, I think it will be alright."

Daya ma gave me some additional insights and advice concerning my personal life and the future, and at the end of our short meeting, she suggested that I could call her secretary anytime I happened to be in Los Angeles, and come to Mt. Washington for a visit.

A Successful Career

I spent twelve (12) years with the USGS, working my way up from hydraulic engineering technician to supervisory research hydrologist. During my career with the USGS, I was often transferred to special projects because of my background in math and physics, and I received additional training in geology, computer programming, systems analysis and mathematical modeling at the University of Arizona in Flagstaff, the University of Iowa in Iowa City, Case Western Reserve, and UCLA. While I was Project Engineer for the cooperative USGS/US Army Corps of Engineers research project, the Hydrologic Engineering Center (HEC) moved to Davis California. It was less than 20 miles, so I commuted. During that time, I took advantage of the fact that the HEC office was adjacent to the UC Davis campus and took graduate courses in environmental engineering and hydrology.

While these things transpired, Dave, after a few years in the SRF monastic order, returned to civilian life, worked for the USGS in Southern California, married, earned a PhD in geophysics, taught at the University of North Carolina, and Southeast Missouri State University, and raised a family. He also became an ordained Methodist minister. Dave and his wife Lee, *another remarkable person*, are well-known by hundreds of thousands of people worldwide for their founding work in the International Association of Parents & Professionals for Safe Alternatives in Childbirth (NAPSAC International) and the Center for Aromatherapy Research and Education (CARE), and Dave is author of a number of important books, some of which have been translated into several languages. Among his books that have proved life-altering for many people, are *Healing Oils of the Bible* and *The*

Chemistry of Essential Oils made Simple, both, in my opinion, classics that will endure for centuries.

Mathematics, Meditation and Metaphysics

When the project in Sacramento was completed, I was transferred to Washington, DC, where I became one of seven charter members of the Department of Interior Systems Analysis Group. While there, I was asked by Self-Realization Fellowship to be Meditation Group leader in DC, where I met several remarkable people. The leader when I arrived there was a man named Gray Ward. He held a PhD in physics from the University of Maryland, and he had accepted a teaching position as associate Professor of physics and electrical engineering at the University of Florida. He confided to me that, in meditation, he had been told that I was coming to replace him. When he left for Florida, SRF moved the meeting place from Maryland to a house on East Capital Street in DC. Among the people meeting there on Sunday afternoons and Thursday nights were professional people, some with degrees in math and science.

One of the SRF Group members was a lady from Australia who also qualifies, in my opinion, as a remarkable person. She was the wife of an Australian Air Force Officer and Diplomat stationed in Washington DC. Sally and I became friends as we lived not far apart in Virginia, and we carpooled to the Meditation Group meetings in DC. I learned that she was able to project her consciousness out of her body during meditation. In fact, she had been able to do that quite naturally all her life. As a small child, she thought everyone did it; but she soon learned that talking about her out-of-body experiences had the effect of scaring people, or causing them to think she was crazy.

While living in Virginia, leading SRF Meditation Group meetings, and working as a member of the USGS Systems Group, I also studied Zen, Tao and Christian mysticism.

As a member of the USGS Systems Analysis Group headquartered in Arlington Virginia, I worked with several scientists at the forefront of applying mathematical modeling techniques to environmental problems, leaders in the state-of-the-art environmental modeling, including Walter Langbein, USGS senior scientist, and IBM mathematician, Benoit Mandelbrot, the inventor of fractals. I had the opportunity to review Mandelbrot's paper introducing iterative equations that produced fractal designs in 1972, three years before it was published in 1975.

Because of the complexity of the environmental problems tackled by the Systems Group, we often had to include stochastic, i.e., statistically derived probabilistic elements, and/or Mandelbrot sets (reiterative fractal elements) in our models. Dr. Mandelbrot, a remarkable man, was often the calm, stabilizing voice in our group discussions. He was a large man, of a very gentle nature, humble, kind and considerate. We were privileged to have his mathematical advice, and his fractal geometry was useful in the modeling of the development and movement of storm cells along cold fronts, coastal morphology and some other natural environmental phenomena. I had no idea how famous he would become.

Most of the scientists I worked with in the System Group held PhDs from Harvard, Stanford and Johns Hopkins Universities. They were all very intelligent, and I learned a lot from them, but I never talked about spiritual matters with any of them except the leader of the group, Nicholas (Nick) Matalas, a

Harvard PhD, with whom I had long intellectual discussions, mostly centered around scientific determinism versus probabilism, mostly related to the modeling of environmental systems. But when we socialized on weekends, we also talked about metaphysical subjects like spiritual evolution and reincarnation.

During this time, I also completed one year of my PhD program in environmental science and engineering at Johns Hopkins University. I commuted from our home in Falls Church Virginia and in order to participate in the Departmental meetings on Wednesday evenings, I found a room at the American University of Metaphysics (AUM) in suburban Baltimore, were there were evening and morning group meditations. After completing one year of my PhD program, the USGS transferred me to Puerto Rico as project engineer of a two-year island-wide water resources modeling project.

Surprising Synchronicity in San Juan

Before moving to Puerto Rico and buying a house in Municipio Guaynabo, I spent a couple of weeks in San Juan, living in El Convento, a hotel in Viejo San Juan (Old San Juan) a few blocks from the Candado Area. The Candado is an area along Avenida Ashford on a narrow strip of land running between the Santurce section of San Juan and Viejo (old) San Juan.

One evening after work, as I was walking along the street enjoying the cool ocean breeze, I heard someone call my name. I looked up and almost directly above my head was a sign reading: 'El Centro de Karma Yoga'. Intrigued, I followed a short path down from the street toward the lagoon on the landward side of the Avenida Ashford. The path ended at a flight of stairs leading to the second floor of a small building sitting on the bank of Laguna Condado. At the top of the stairs was small porch-like room overlooking the lagoon. Windows were open, allowing the aroma of flowering Flamboyan and Flor de Maga trees to waft through the room.

Several pairs of shoes and sandals were lined up along the wall, so I removed my shoes and sat on a bench facing the door at the opposite end of the room. The door was obscured by strings of colored beads hanging from the ceiling almost to the floor. I could hear soft voices coming from inside but couldn't make out any words. I planned to wait until the class, or whatever it was, broke up to talk to someone about meditation meetings. I had been the SRF Meditation Group leader in Washington DC for about three years, and it would be nice to be part of Yoga meditation group while working and living here.

I was sitting there quietly, with closed eyes, when I heard the beads stir and a small dark-eyed woman with black hair done up in a large knot on top of her head, poked her face through the beads and beckoned to me, saying:

"¡Adalante! ¡Estamos esperando!" (Come on in! We're waiting!)

I followed her into the inner room where several people were sitting on pillows in a circle on the floor. She said something to the group in Spanish, that I didn't catch, and then turned to me and said in English: "I am Aury Moya. Please explain to us how you meditate."

I continued to attend meetings and at the request of Aury, the founder and Director of El Centro de Karma Yoga, I taught weekly classes in meditation techniques there. After a few months, friendships developed with several of the Karma Yoga

Center students, and one of them, Carlos, invited several of us to his cabin, up in the El Yunque Rain Forrest one weekend for a picnic and swimming in a mountain stream which was called El Rio Sangre de Los Santos. (Blood of Saints River). Conversation got around to that first night when I climbed the stairs to the Yoga Center, and I asked about the strange way Aury had invited me into the meeting. I mentioned that I thought that she may have been expecting someone else, and I just happened to show up. Carlos looked at me quizzically, smiled and said:

"No, man. She told us for several weeks, that when the class got to the point of discussing meditation techniques, someone from the States would be here to teach that part of the course for her. The week before you showed up, she said 'next week will be the meditation class'."

"But, I had never met Aury, or talked with her, before that night." I protested. "How could she have known that I would be there that particular night?"

Carlos laughed, and said: "Aury knows a lot of things, man! She described you to a 'T', and even said your name was Eduardo! And, when we were all there that evening, we asked 'where is that meditation teacher?' And she said: Be patient, 'he'll be here in a few minutes' - and then you showed up!"

I think that the experiences I've described so far constitute pretty good evidence that synchronicities and pointed clues have guided me to a degree of alignment that has resulted in a remarkably blessed and ordered life, including experiences of a spiritual nature that cannot be explained within the simple materialistic paradigm of mainstream science. Analogous to following the clues of recurring patterns while solving the cube, by recognizing synchronistic patterns in my life, I have been guided to increasingly higher levels of alignment and understanding. More and more often, it is just a matter of going with the flow.

Trouble in Paradise

But it has not always been easy. I cannot say that I have always been aligned with the laws of the universe and have never made any wrong decisions. That would be wonderful if it were true, but it is not. At times, I have stumbled, blind to the signs, and synchronicities provided by the universe. Sometimes I had to act, even when the clues are confusing, and I was unable to see the whole picture. That is what happened to me in 1977. My life became intolerable, and I decided to leave my job and end my marriage.

I gave my two-week notice, and I left nearly everything I had accumulated behind. I signed everything over to my wife except for one of our three cars, which I kept, to go on the book tour. I left Tampa with nothing but a few personal belongings, a change of clothes and my severance pay. I did not mind giving up *things*, but the emotional results of leaving my marriage were devastating, for me and my family.

I had met Jacqui about a year before I left the USGS, but just considered her to be a special friend on the spiritual path. I knew there was a deep connection there, but during my troubled years with Bobbi, my first wife, I often said that if this marriage should come to an end, I would never marry again. I planned to apply for acceptance into the monastic life in Self Realization Fellowship. A senior monk had assured me that because of my sincerity and serious study and practice of SRF meditation techniques, if I were not married, I would most likely be accepted, Destiny, however, cannot be denied. Jacqui and I were married the day after my divorce was final (we will share more about this later in the book) and we soon had a son. We named him Joshua.

My first book was not exactly a best seller, so I had to go back to work. After a few part-time jobs, I landed a job as an environmental engineer with R.M. Parsons in Pasadena, California and Jacqui got a job as an assistant to a vice-president of First Interstate Bank in Downtown LA. We began to think about saving money for a house, but I was sending monthly child-support checks to Bobbi and offered to take one or both of the children. Bobbi refused to let me have custody of the children, but did allow them to visit a few times. However, none of the visits worked out very well. Living in California was expensive, and we were not able to save much. So, when an overseas job that paid twice as much as I was making in California came available. I saw it as a chance to make enough money to be able to save to buy a house, and Jacqui and Joshua could go with me, so I applied and got the job.

Jacqui was not thrilled about going to Saudi Arabia, even though it would just be an 18-month contract. When I mentioned that we could visit India on the first month-long break, she agreed. But I knew she did not like leaving California, and the contract in Saudi Arabia was not an easy one. I was the first environmental engineer for the Industrial City of Yanbu, a refinery and deep-water port being built at the west end of the Trans-Arabian Pipeline on the Red Sea, about halfway between Jeddah and the Holy land. One of the engineers at Parsons told me that 18 months in Saudi Arabia was a real test for any married couple. Was I about to make the same mistake I had made with my first marriage?

It was difficult. I had to work ten-hour days, six days a week. And Jacqui could go nowhere without me. But we survived it, and I was able to double my child-support checks, and we were still able to save money to buy a home. My dad found a house for sale in a little town on Black River in Southern Missouri. It was a place virtually on the river bank and life in a small town in my native Ozarks seemed like heaven after the hustle and bustle of LA and the stressful 18 months in Saudi Arabia. We loved it, especially Joshua and me. I got a job teaching high school math in a nearby town, and we enjoyed fishing and swimming in the river in the summer, but after about half-way into the second year, Jacqui was offered her old job back at First Interstate with a promotion and a raise. I was tired of watching the grass grow, so we put the house up for rent and moved back to LA.

Opportunity and Tragedy

After about a year back in LA, I got an unexpected offer of a high-paying job in Saudi Arabia with another company, located in Jeddah, instead of a construction camp in the middle of nowhere. So, in 1985, we returned to the Middle East for a two-year contract. The job was Marketing Director and Hydrogeologic Consultant for Amartech, a company based in Concord MA. Within a few months, I obtained a huge contract with MEPA, the Saudi Military EPA, for installing and operating a kingdom-wide environmental monitoring system. It was the biggest job Amartech had ever landed, and I was hailed by the company as a hero.

We rented four floors of a major office building in Jeddah and began buying computers and other electronic equipment and recruiting and hiring staff. But, just as things seemed to be going great, tragedy struck. Not one tragedy, but two devastating tragedies struck: The Saudi Government froze all contracts over 120 million dollars, and Jacqui was diagnosed with cancer. She had one operation in Jeddah and then we flew back to Massachusetts General Hospital in Boston, where she underwent another operation, was treated with what was then an experimental procedure: linear radiation, and was given two months to live.

Everything is Connected and the Power of Prayer

You may be wondering what all this has to do with the cube. Be patient. What I wish to show is how all things are connected. Remember this old spiritual?

Dem bones, dem bones, dem dry bones!
Dem bones, dem bones, dem dry bones!
Dem bones, dem bones, dem dry bones!
Now hear the sayin's of de Lord!

De foot bone connected to de leg bone,
de legbone connected to de knee bone,
de knee bone connected to de thigh bone,
de thigh bone connected to de hip bone, ...
Now hear de sayin's of de Lord!

So goes the "Dry Bones" Song.

But, it's not just the bones, *everything* is connected. The substance of the physical world, the body, bones, blood, brain, and all; all are connected to the atom, the atom is made of the proton and the neutron, connected by the electron, the proton is made of the up-quark and the down-quark, the electrons and quarks and everything are made of quanta of mass energy and gimmel, **and gimmel is the third form, the result of the AUM, the word of the Lord!**

Gimmel is the real connection between the physical world, individualized consciousness and the Primary Consciousness of the universe. Once you realize the truth of this, you are no longer at the mercy of random events. You can see, feel and experience the connection of all things.

In 1985, when Jacqui was diagnosed with terminal cancer, we refused to accept the death sentence. While Jacqui was in the hospital in Boston, I scoured the reference books in the libraries for information about her cancer. I learned that cancer is not just one simple thing. I learned that medical science had identified more than 200 kinds of cancer, and that many of them progress very differently from one patient to the next. I also learned that chemotherapy is prescribed virtually 100% of the time for advanced cases of cancer, while it is only effective about 3% to 4% of the time. Not only that, patients often died from the treatment, rather than from the cancer. In fact, the chief oncologist at Massachusetts General told us that, if Jacqui didn't die from the cancer, she would die from the treatment! He firmly announced to us that she had approximately two months to live.

I am not quick to anger, but I confronted him, and I did not mince words. I said:

"You should not be telling my wife that she is going to die in two months.! You have no right to pronounce a death sentence on anyone!"

The doctor reacted as if I had slapped him in the face. He replied:

"That's what all the studies show. Statistically, a person with your wife's diagnosis will die within two months!"

I looked him straight in the eyes and declared: "Jacqui will not be a statistic!"

Jacqui refused chemotherapy. We did accept linear radiation, which was experimental at that time. There were only two places in the country where it was being used: at Stanford University in California, and at Massachusetts General. But the prognosis was the same: they thought that if she didn't die from the cancer, she would die from the treatment.

We turned to prayer. We meditated and prayed at least an hour every morning and evening, and our friends in California put her on the prayer list of the Worldwide Prayer Center at SRF Mother Center in LA. Friends around the world prayed for her recovery. Jacqui "miracuously" recovered and was still very much alive, more than 30 years later.

Focused intent can establish direct connection with Primary Consciousness through the consciousness of *gimmel* in every atom of every cell in your body. No one needs to die because of the side effects of drugs. Does this mean that everyone can forsake medical treatment and heal themselves? No. Medical science does save lives. Illness and death are part of life, and everyone must make his or her own decisions to the best of their ability, based on their individual knowledge and belief.

But, don't forget, knowledge trumps belief every time. If you know exactly how and why something works, it *will* work. But it works not just because of physical law, but because of spiritual law, as well.

The good news was that Jacqui was free of cancer. The bad news was that, because the cancer was diagnosed outside the country, our insurance company refused to pay the massive medical bills. They paid only one thousand dollars of a huge medical bill. We returned to the US, after having worked three and one-half years in Saudi Arabia, with our earnings wiped out and deep in debt. But - we survived it!

This experience in 1985 and many experiences since, led us to explore the use of alternative treatment modalities, including chiropractic, acupuncture, herbs, essential oils, and prayer in conjunction with, and sometimes instead of allopathic medicine. Life is a precious gift and you need all the longevity you can get in order to fulfill the purpose of your life and achieve your highest potential.

This book was not intended to be an autobiography, so I will skip over several years of experience and learning, to some of the more important examples of synchronicity and alignment. An important part of our journey has been learning about the natural medicines of Biblical times: essential oils. We were introduced to essential oils in 1999 by my life-long friend and very remarkable person, Dr. David Stewart, and that led to a new period of learning, prosperity, and meeting more remarkable people.

A New Lease on Life

We moved from California back to Southern Missouri in 1992 for several reasons. My dad had passed away, my mother was in her mid-70's and I was their only child. We were also concerned about the deterioration of California schools. California once had the best public schools in the country. But. when we returned from Saudi Arabia the second time, Joshua had been in a British Dutch school in Jeddah, and we found that he was years ahead of the curriculum in California.

The public schools in California had become bogged down in social engineering and were riddled with gangs and drugs. We couldn't afford private schools in California and they didn't seem to be much better, anyway. The schools were better in Missouri, and much less dangerous, but they had not kept up with population growth, so class rooms were crowded, and Josh was still ahead of the curriculum and consequently very bored.

Jacqui had agreed to move back to Missouri, under one condition: we would return to LA every year for the SRF convocation as long as we were able. I agreed, and this we did for twenty years. Life improved for us in Missouri. My qualifications and experience as an environmental engineer, meant that I was in demand by engineering firms, and I found a job in Cape Girardeau, Missouri, in a small branch office of Dames & Moore, a large engineering consulting firm with one of their major offices in St. Louis. This meant we could afford a nice home on a few acres at the edge of town, something that had been completely out of reach for us in California.

Another benefit of moving back to the Midwest in 1992, especially for Joshua and me, was the opportunity to reconnect with nature in the Southern Missouri Ozarks. My favorite spot on the edge of the Current River National Scenic Riverway, was virtually unchanged since my high-school days in the early 1950s. Joshua and I enjoyed week-long hiking, camping, fishing and swimming visits, and even camped in the same cave my college roommate and I camped in more than 40 years earlier. It had been called Buzzard Cave when I was a teenager, Honeymoon Cave after Dave and his bride spent part of their honeymoon there, and now it was called Ride-in Cave by Chuck Golden who had developed a trail-ride resort known as Golden Hills in the area. We became good friends with Chuck and Kay Golden, who were truly remarkable people. They had lived the American dream, beginning as children of poor hill folk in the 1930s, to becoming self-made millionaires by 1978. Chuck explained how they did it in his 2007 book: *Building and Living the American Dream, how to get from where you are now to where you want to go.* Chuck and Kay achieved their goals by putting God first, setting goals and working tirelessly to serve others. Their success is a good example of the process modeled by the solving of the sacred cube,

Chuck decided that unloading bags of cement and picking cotton, for $2 to $4 per day, was not the way he wanted to spend his life. He had an early interest in physical fitness and naturally avoided things that most of his friends and family indulged in, like smoking, drinking alcohol, and even coffee, because he knew they would impact his health. As a member of his high school wrestling team, he began to dream of becoming a professional wrestler. He developed a healthy life style, because he dreamed of becoming a professional wrestler. When Chuck graduated from high school in 1955, he couldn't afford college, so he enlisted in the Air Force in a program that would allow him to go to college and get an education. He played football while in the Air Force and focused on physical fitness so that he could pursue his dream of becoming a professional wrestler after his stint in the Air Force.

He met Kay Hodges, the love of his life, before achieving success as a professional wrestler, and they began to plan to achieve worthwhile lifetime goals together. After their marriage separately and together, they achieved some success and recognition: Chuck as a professional wrestler (billed as a good guy, the Golden Boy), and Kay as a spokesperson for General Electric appliances on National TV. But these successes involved them being apart too much. Chuck and Kay wanted a life together, so one weekend, they sat down together to discuss what they could do *together*, that they could both enjoy for a lifetime. They agreed that they would enjoy owning and working in a health club. They agreed that Chuck would quit professional wrestling and they moved to Mesa Arizona and opened Golden's Health and Racquet Club.

Like most young couples starting out, they had their successes and failures; but different than most, they had a plan. They set a goal to be millionaires by the time they were 40. Their plan involved deciding what they wanted out of life, and setting goals to achieve in six major areas of life. Here they are, presumably in order of importance to Chuck, taken directly from Chuck's book:

Spiritual (Faith)
Family (My wife, children and grandchildren)
Physical (To be in excellent health)
Education (Knowledge that I need to take me to my goals and through my life)

Social (To be able to communicate well with others)
Financial (To have the money I want to do what I want with whom I want)

By not being afraid of hard work and long hours, Chuck and Kay kept moving toward their goals. They read and listened to motivational writers like Napoleon Hill, Norman Vincent Peale and Tony Robbins, and applied the power of positive thinking, avoiding negative thinking and negative people, like many friends and family who said "You'll never reach those lofty goals", and when they did achieve something good, said "You were just lucky." They did indeed achieve their goals. They were millionaires by the age of 40, and in 1978, they sold their businesses and buildings in Arizona for two million dollars. After more than 40 years in the health club business, they decided to pursue a new dream: to own and operate a ranch and trail-ride business in Texas County Missouri. That ranch just happened to include much of the Big Creek Basin on the western edge of the Current River Basin that I had explored as a teenager, and the cave where Dave and I had camped in 1958. When Chuck learned that I was familiar with the area, and was already a published author, he asked me to write a trail-guide book. I readily agreed because it gave me an excuse to spend some time in the wilderness area I loved so much. In 2003, I published the book Big Creek, History, Folklore and Trail Guide, which was immediately popular with trail ride people, and is still available at Golden Hills Ranch and Resort.

New Challenges and New Solutions

After a few years back in Missouri, Dames & Moore wanted to move me to a larger office in Chicago. My mom was in assisted living by this time, and we couldn't move her, and we had no desire to move to Chicago. The winters were bad enough in Southern Missouri, and we had lived in large cities most of our lives together and were enjoying a quieter, calmer environment with the wonderful proximity to relatively pristine nature. Not only were we within a few hours' drive of Big Creek, wild turkeys, deer, foxes, bob cats, rabbits, and squirrels frequented our backyard, and birds of nearly every kind flock to our feeders. We were located in the Mississippi River Valley, on major bird migration paths, and the area is home to many year-round resident species. We could not and would not move; so, we had to find a "plan B".

In 1995, we rented a small office on the square in Jackson, Missouri, the county seat of Cape Girardeau County, and opened an environmental consulting business. My engineering colleagues in St. Louis were of the opinion that we would probably fail in this location. Cape Girardeau County had a population of about 50,000. The reason they closed the branch office where I was hired, was insufficient work. And it was hard. We worked 10 and 12-hour days and 60-plus hour weeks. With my PhD and years of experience, I was able to find jobs in the four-state area. We were located in Southeast Missouri, within an hour's drive of Southern Illinois, Kentucky and Arkansas. Jacqui, with her background in office administration and accounting, took care of the administrative tasks, and I did the technical work. We made a good team.

Again, we survived, and by 2000, our consulting business had grown from the two of us to about 15, an administrative assistant to work with Jacqui, and geologists, engineers and technicians under my direction. We had annual project values of about one-half million dollars. But, with the long hours and the stress of being responsible for insurance, payroll and taxes, in the hot and humid summers, cold wet winters of Southeast Missouri, Jacqui began to suffer from headaches and the aches and pains of sciatica and fibromyalgia. Essential oils proved to bring relief.

At my insistence, Jacqui went to a Young Living meeting in Toronto with Dave and Lee Stewart and received training from the founder of Young Essential Oils, Gary Young. She studied, took the exam and became a registered aromatherapist with the National Association for Holistic Aromatherapy (NAHA). She also became a certified instructor for the Center for Aromatherapy Research and Education (CARE) an organization founded by Dave and Lee Stewart.

The stress of working long hours, exposure to environmental toxins and weather of all sorts at job sites had taken a toll on my health as well. I was experiencing respiratory problems, dizziness, asymmetric weakness, and fatigue. When I went to see a doctor, he suggested that the problems were simply due to hard work and age. I did not accept that diagnosis. In 2002, we went to the Young Living Clinic in Utah for a week-long program of treatment. By the middle of the week, I felt so much better, I couldn't believe it! Jacqui's health was greatly improved also. We began using Young Living essential oils and supplements, as well as all of Young Living's toxic-chemical-free products, and we enjoyed excellent health again!

A Couple with a Great Vision: Gary and Mary Young

In 2006, as a result of a discovery Jacqui and I made in the course of our work as environmental consultants, I was invited to speak at that year's Young Living World Convention in Salt Lake City Utah, where I presented experimental evidence from a field study I had conducted showing that mold infestations, including species that produced toxins harmful to human health, were effectively eliminated by diffusing a non-toxic blend of natural essential oils. This led to getting to know Gary Young and his wife, Mary, visionary founders of the Young Living Essential Oils Corporation, two truly remarkable people.

As it happened, about the same time I was conducting the field studies, their youngest son, Josef, had been exposed to toxic mold in the basement of their home in Utah. He had been hospitalized and was near death. The Youngs were well aware, through extensive research, both in their own labs and in independent labs, that several essential oils and blends of essential oils are effective anti-fungal agents, and using them, they were able to save their son's life, after the doctors had given up and released Josef from the hospital into their care. But, to their knowledge and mine, no one had designed a protocol and conducted field tests with essential oils before. Soon after that, Jacqui and I wrote <u>Nature's Mold RX, the non-toxic solution to toxic mold</u>, and began to help the Youngs in their visionary quest to put therapeutic-grade essential oils into every home, work place and public building around the world.

I first met Gary Young at the 2006 Convention as I was coming off the stage after completing my presentation with Power Point slides of my research proving that non-toxic essential oils are effective in eliminating mold from home and work environments. It was one of Young Living's five-oil blends that proved to be the most effective among all the oils we researched and tried. We shook hands and became friends. This meeting led to a professional and personal relationship with Gary and Mary Young that has continued until this day. Jacqui and I have been invited speakers at numerous Young Living meetings across the country and in Australia.

The Authors, Speakers at a Convention in 2013

Intellectual Synchronicity: Dr. Vernon Neppe

In 2008, Jacqui gave me a book titled *"Thinking on the Edge"*, a collection of essays by members of the International Society for Philosophical Enquiry (ISPE), a society for individuals with high IQs interested in creativity and longevity. I was impressed by a couple of the essays and decided to see if I could become a member. The entry level was nearly 20 points higher than the entry level for MENSA, so I wasn't sure I could qualify, but my high school counselor had said that I could succeed at anything I wanted to do, so why not try?

I took their test and qualified. I joined and quickly advanced from Associate to Full Member, to Fellow, to Senior Research Fellow, and Diplomate. More importantly, I met Dr. Vernon

Neppe, MD, PhD. a member and world-renowned neuroscientist. We soon began to work together on theories of common interest. We coordinated our research by email, and Skype once a week. We were both excited about our common interests and the possibilities of a scientific breakthrough, a paradigm shift that could change the world.

As our work intensified, in early February of 2010, I let Vernon know that I would be out of touch for several weeks because I would be joining Gary and Mary Young, a film crew and a group of Young Living distributors in the Middle East to help make a film about the Frankincense Trail from Southern Arabia to the Middle East and Egypt. Vernon responded that it would not be a problem because he and his wife would be going back home to South Africa during February and March, visiting relatives in Europe on the way. They had already made their flight and hotel reservations, and when I called to tell Vernon of my trip, Young Living had already made my reservations and I had the tickets in hand. Knowing that flights to Africa usually involve stop-overs and plane changes in Europe, Vernon asked where I would be changing planes. It happened that we would both be in Amsterdam on the same day! Synchronicity!

Even though we felt like old friends, Vernon, living in Seattle, and I, living in Missouri, had not yet met in person. That Saturday, I had a nine-hour layover in Amsterdam, and Vernon and his wife Lis were staying in a hotel in downtown Amsterdam. When I landed at Schiphol, I exchanged some US dollars for guilders, hopped on the public transit and arrived at their hotel within an hour. We spent several hours talking and Vernon recorded our conversation as we discussed our ideas and I explained how I could prove why up- and down-quarks must combine in threes to form protons and neutrons using Fermat's Last Theorem.

Vernon Neppe, MD, PhD. FRSSAf, DFAPA, BN&NP, FFPsych, MMed, DPsM, DSPE*, is one of the most renowned, successful and gifted neuroscientists in the world today. The interesting thing is, we both felt that we were destined to work together as soon as we became acquainted through electronic media. Once we started working together on a comprehensive theory in earnest, we made rapid progress, producing the Triadic Dimensional Paradigm, a shift to a consciousness-based scientific theory that really works. In 2011, we published numerous papers and the book *Reality Begins with Consciousness*. Professor Neppe certainly qualifies as one of the most remarkable people I have ever met.

INTERNATIONAL SCIENCE AND SPIRITUALITY CONFERENCE (L to R) Vernon Neppe, MD, PhD, Edward Close, PhD, Maria Sagi, PhD, and Erwin Lazlo, PhD

*See more about Dr. Neppe in Notes and References at the end of the book.

Remarkable Experiences in the Middle East

I jumped at the chance to be part of the Young Living Frankincense Trail Expedition, because I'd wanted to visit Petra, the fabled city in stone, the city of the Nabateans, in Southern Jordan, ever since seeing the Indiana Jones movie featuring part of it. I had once been only about one hundred miles from Petra, when I worked and lived in the Middle East in the 1880s. And, although I had been to Cairo, I had never been to the Pyramids, Luxor, Abydos, or the Valley of the Kings. I had an intuition that this would be the trip of a lifetime. I could be helpful to the expedition because of my knowledge of both Eastern and Egyptian Arabic and my experience in the Middle East.

On the flight from Cairo to Luxor, Dr. Young came back to where I was seated and asked me if I would consider playing the part of the physician-priest of the caravan in the Frankincense film. Of course, I was thrilled and honored. I would be riding second camel, between two assistants, behind several horsemen headed by Dr. Young as the leader of the caravan. The caravan consisted of five horsemen and seventy-five camels, stretching out across the desert. We filmed the start of the caravan at a reconstructed village south of Giza and in the desert near the "bent pyramid" and night scenes at a desert encampment, after a staged attack by bandits. We would all be in authentic costume.

The Ancient City of Petra was a lost city, hidden in the mountains of southern Jordan for thousands of years before the Swiss explorer Johan Burckhardt found it in 1812. Petra, like the Great Pyramid, is one of the Seven Wonders of the Ancient World. This entire city of temples, tombs, and residential buildings was carved out of multi-colored sandstone cliffs. Its mystic grandeur cannot be captured on film; it is like nothing else anywhere on the planet. The largest of the buildings, called The Treasury, was used as a safe holding place for the precious frankincense and myrrh resins brought nearly 1,500 miles by camel caravan from trees growing in Southern Arabia, in what is now Yemen and Oman, to Petra, Jerusalem, Damascus and Egypt, where the kings and other elite used them for incense and medicines.

An Unusual Experience in the Great Pyramid

Between the days of filming in Egypt, we were able to visit most of the temples, tombs and pyramids of ancient Egypt. Of special interest here, is one afternoon in the Great Pyramid of Giza: A few minutes before four o'clock in the afternoon on Saturday, February 27th, 2010, we hiked up the hill from the Sphinx to the Museum housing Khufu's Solar Boat, the boat said to be built for the purpose of carrying the Pharaoh Khufu to the afterlife. This is a one hundred and forty-three-foot long cedar wood boat constructed at or before the time of Khufu's death in 2566 BC, or at least 4580 years ago. It had been unearthed and reconstructed only a few months before our arrival there. But, unfortunately the newly constructed building housing it was closed, so we continued on up the hill to the Great Pyramid.

I was feeling great, and in good spirits as we prepared to enter the Great Pyramid. Dr. Young said that he could not be held responsible for what we might experience in the pyramid, but he would like to hear about the experiences. I had no feeling of expectancy, however, and was fully in the moment, inspecting the huge stone blocks of the pyramid, noticing three large black birds circling around the apex of the Great Pyramid, and enjoying the camaraderie of the group of Young Living people. We entered the pyramid through a rough-walled tunnel, known as the Robber's Tunnel, blasted through the ancient masonry about 820 AD. I remember thinking as we entered that it was a shame that the Great Pyramid had been violated in this way.

Preparing to Enter the Great Pyramid

I was in the second group of about twenty people to enter the Pyramid. The first group went up the Grand Gallery to the King's Chamber, and the group I was in took the lower, nearly horizontal passageway to the Queen's Chamber. The passage was only a little over three feet high, with a step down near the end. After entering the chamber, I crossed the room and stood near the corner opposite the entryway. We were told that the lights would be turned off for a few minutes so that we could experience total blackness. Shortly after the lights went out, someone started chanting "OM" over and over. I joined in, and soon it sounded as if everyone had joined in. The effect was very peaceful and uplifting.

Because I was an experienced spelunker, and had explored a lot of caves in my native state of Missouri, I knew how dark it can be in a rock-bound chamber. It was pitch black. I think I had my eyes closed most of the time, but a few seconds before the lights came back on, I saw a ball of light coming straight at me from a point near the ceiling in the opposite corner, above and to the left of the alcove in the east end of the chamber. The ball of light was white, a little larger than a softball, with smaller lights spinning around it. I didn't know it then, but others in the chamber saw it too. If they hadn't told me about it later, I would have thought that the experience was entirely subjective.

The ball of light struck me on the forehead, but there was no feeling of impact. Instead, I immediately experienced extreme vertigo. I could feel the spinning of the planet on its axis. I could also feel the rushing of the Earth through space in its orbit around the Sun, the Sun moving through the Milky-Way Galaxy, and the Galaxy wheeling its way through the universe. I could feel all these spinning motions at once! There was no up or down, only the sensation of a swiftly accelerating whirling motion. I remember saying: "I can't stand up!" and someone on either side of me grabbed my arms to keep me from falling to the stone floor. We struggled to the low passageway, and about halfway along it, I became violently ill, threw up and left the body. This was a different kind of OBE (out-of-body experience), and it wasn't at all pleasant!

I was in and out of bodily consciousness several times after that, as our tour guide Mr. El Komaty and one of his men drug me out of the narrow passageway, and literally carried me out of the Pyramid. I remember speaking to them in Arabic and thanking them. They placed my body on the stone blocks at the "Robber's Entrance" and several Young Living healers began working on me. What was really unusual about this experience, was what happened between episodes of spiraling vertigo and violent regurgitation.

After the ball of light struck me and the world went spinning and spiraling out of control, I began to see lights, hear sounds and see symbols and faces. When the spinning got too intense to bear, I left the body. While out of the body, I felt no spinning, no vertigo, and the lights, sounds, symbols and faces became clearer, but, when I turned back toward my body, I began to experience extreme vertigo again, increasing to the point that I was forced to throw up. Each time I threw up, I was propelled out of the body again. While out of the body, I experienced a series of vivid dreams or visions.

During one of them, I suddenly saw the Great Pyramid from above. The first thing that struck me was that there were no buses in the parking area, and there were lots of green grass and trees! Then I saw that the Pyramid was covered with smooth white stone, and three black birds, like those I had seen before entering the Pyramid, were circling around a burnished gold cap on the top of the pyramid, shining in the sun. When I opened my eyes again, the world was still spinning. Dietfried, an Austrian friend in the group handed me some dried Japanese seaweed which I ate. It seemed to help.

I vaguely remember being carried down the side of the Pyramid and across the flat area to the bus. Two members of the Young Living group, a doctor and a nurse, accompanied me. Their attention, positive energies and support were very comforting and very much appreciated. During the bus ride to the Cascade Hotel, where we were staying, every time I closed my eyes, I felt increasing vertigo and saw a stream of faces, lights, sounds and symbols again. At the hotel, I still couldn't stand, so they helped me into a golf-cart and whisked me to my room, where they deposited me on the bed. Tom Reed, who was my assistant in the film, was actually a doctor. He took my pulse, and helped me to drink electrolytes and some Ninxia-Red Juice, while the nurse, Brenda Charbonneau, applied energy healing techniques, after which she and Mr. Ashraf, one of our Egyptian tour guides, went to the hotel kitchen to get some hot chicken soup, a little of which I managed to eat before falling asleep. While asleep, the stream of faces, lights and sounds continued like a dream that would not go away, and for the next day and a half, every time I closed my eyes, the vertigo returned, and the streaming images continued.

The "Treasury" in the Ancient City of Petra

Petra, The Ancient City in Stone

The very next day, we flew to Amman Jordan and then traveled by bus the 140 miles to Petra. During all this time, I worked hard to stay fully awake, almost propping my eyes open to avoid the possibility of vertigo and vomiting. About half-way through the bus trip from Amman to Petra, we stopped at a restaurant for lunch. I avoided eating and looked around the gift shop while the others ate. When we arrived at the hotel just outside the entrance to the Ancient City of Petra and checked in, by the time we had taken our luggage to our rooms, the sun was only an hour or so above the western horizon.

Still not feeling fully recovered from the ordeal that started with the ball of light in the great Pyramid, I decided to forego a trip into modern Petra, or a short tour into *Al Siq* (the narrow canyon leading into the Ancient City) and stay behind in the hotel. I decided that I should call Jacqui because, if someone in the expedition had called friends or family in the US and mentioned that something had happened to me in the Pyramid, she might be worried. It was something that at least all the people who were part of the group in the Queen's Chamber with me knew about, and many of the Young Living "family" knew Jacqui. I was worried that someone might call her with partial, or maybe even unintentionally exaggerated or distorted information, and she would be worrying unnecessarily.

When I reached her, I learned that she had experienced some of the same things I had, at the same time, while at home more than 6,400 miles away! And she knew that I was OK. From a scientific point of view, this only increased the anomalous nature of the experience and further challenged me to try to find an explanation for it and understand exactly what had happened, as well as why and how it had happened.

Happy that Jacqui was OK and not worried, after I ate a light snack in the hotel restaurant, I strolled alone outside the hotel grounds, turning left on a sidewalk, away from the Modern City of Petra. After a very short walk, the sidewalk ended, and I was on the edge of Petra, looking north and west over the spectacular sandstone mountains that I knew concealed the Ancient City. The sun was just above the horizon.

I sat down on the ground to rest and noticed a piece of white and tan alabaster about the size of my fist on the ground beside me. When I picked it up, everything suddenly changed. I was a young boy, running toward the Siq, with the chunk of alabaster in my hand. I could hear camels grunting and the padded sound of their feet on the sand and gravel, as an incense caravan was approaching. I dropped the piece of stone and the vision faded. I returned to the hotel and to my room to rest, a little shaken and bewildered by what had just happened.

What was I supposed to learn from these experiences? It certainly was further evidence that consciousness is not just a product of chemistry and electrical impulses in the physical body, but I already knew that, so why did I have to go through these horrible and upsetting experiences? Mary Hardy, another Young Living member and author who writes about pyramid energy, helped me understand later, that it was about *alignment*. In the pyramid, I was facing east, the direction toward which the Earth spins, and the ball of light came out of the northeast corner of the chamber.

According to Mary, I was part of a shifting of the alignment of the Earth and the Cosmos, and the stream of faces, voices and symbols was a download of information stored in the Great Pyramid; and then something similar had happened the next day, when I received information stored in the stones at Petra, as it was directly downloaded into my brain.

Clearly, synchronicity, intuition and alignment were all at work throughout this trip.

Synchronicity occurred when, without any planning on our part, Vernon and I were in Amsterdam at the same time (See **Part II** and **Part V**). Intuition played a role in my knowing that I was supposed to be on this trip, alignment happened in the Great Pyramid, and I seemed to be seeing for a brief moment, into the past, both at Giza and in Petra.

Exactly what was in the images and information downloaded from the Pyramid? The short answer is a lot! But the long answer is far beyond the scope of this book. I can tell you that, based on some of the information I received, the history of our planet and the universe we live in may be very much different than what is deduced and projected by the investigations and findings of mainstream anthropologists, geologists and cosmologists.

This is not meant as a criticism. I am a scientist, and I know that scientists are human. Scientists make projections based on the information available to them, prevailing theories and their own personal beliefs. For starters, as the cube and quantum physics indicate, matter, energy, time and space are not what they appear to be. This means that any history that assumes that space and time are uniform and linear, providing a uniform objective domain within which all things happen, is likely to be distorted, and maybe even completely wrong.

The Author as Physician/Priest with the Caravan as it Arrives in the Ancient City of Petra

After filming in the Ancient City of Petra, a few of us, whose flights did not leave Amman until the next day, were treated with a trip to a Greek Orthodox Church in Madaba, to see one of the oldest maps made of ceramic tile in the Middle East; on to Mt. Nebo, where Moses died at the age of 120 years; and then to the River Jordan, where John the Baptist baptized Jesus, and finally to the Dead Sea, near the mouth of the Jordan River, before returning to Amman.

PART V: THE ULTIMATE SOLUTION 175

Mt. Nebo, Monument to Moses

On the River Jordan where John Baptized Jesus

During this tour, on the east bank of the Jordan River, looking across into Israel, seeing the blue and white Israeli flag flying such a short distance away, I had another unusual synchronistic moment: I was overwhelmed with emotion for no apparent reason. I stood looking across the River Jordan, with tears running down my cheeks, and a strong feeling of *déjà vu*. I had no distinct memory of a past life here, or vision of a different time, but something was pulling at my heart. Several years later, with a DNA test, I found that I had ancestors, several generations back, who lived in the Middle East, Northern Egypt, and, yes, there was a Jewish ancestor. Did this emotion come from my DNA, a somatic memory, or was there a deep memory of a past life in the Holy land? Perhaps.

Israel on the Bank of the River Jordan

PART VI: PATTERNS OF GOOD AND EVIL

Part of a Greater Reality

What does all this mean? Have I had contact with enlightened beings during the journey of my life? Yes, I think so. It appears that there are very few who have finally and completely realized their Eternal Oneness with Primary Cosmic Consciousness, perhaps only one or two, but we are all somewhere on the journey between complete confusion and disorder, like the scrambled cube, and complete coherence. Am I myself a fully enlightened being, filled with the glory of Cosmic Consciousness? No, I don't think so. But I believe it is possible for each and every one of us to attain Cosmic Consciousness, and I am working on it every day to the best of my ability.

There is strong evidence that the physical universe is only a small part of a much greater reality that exists as reflections or manifestations of the patterns of information and awareness that exist in the Conscious substrate of that Greater Reality. The physical universe that is available for sentient beings (that's us) to experience thru the physical senses and extensions of them in 3 dimensions of space and one moment in time, exists as a vast number of dynamic distinctions of mass and energy, continually changing in accordance with a set of specific mathematical relationships that convey the patterns of the Conscious Substrate into the physical domain for the purpose of creating stable structures of quantum distinctions capable of supporting organic life forms that act as vehicles through which individualized consciousness, i.e., sentient beings can experience reality and grow in awareness and understanding.

When an individual sentient being masters *all* of the aspects of reality that a physical being is capable of experiencing as real, in terms of distinctions of mass-energy in space-time, that being advances toward the ultimate goal of all sentient beings, Cosmic Consciousness, and the fine tuning of the parameters of the universe supporting the physical evolution of life and consciousness make perfect sense. It is clear to me that the conservation of mass, energy and gimmel and the quantum continuity of reality implies continuation of individual consciousness in some form, and individual spiritual progress made in any life is never lost. But, if the reason for existence is for each and every one of us to achieve Cosmic Consciousness, is one life enough? From the evidence of millions, it doesn't appear that many make it in one lifetime.

Is Reincarnation Part of Cosmic Recurrence?

When I had finished the first complete draft of this book, I sat down and relaxed. I picked up a cube and scrambled it. I solved it, and scrambled it again, solved it and scrambled it, solved it, scrambled it, and solved it again. Then I noticed something interesting: The way I was solving the cube was subtly changing! The methods of solution presented in this book places the primary focus and concentration on the final stages of the solution. The complexity of the algorithms and therefore the concentration and focus, increases from the first step, i.e., from the completion of the red cross, to the completion of the first, second and third layers of the cube. And we've discovered the analogy of the first, second and third layers to the alignment of body, mind and soul, respectively. While repeatedly solving the cube over and over again, however, my focus was changing! I began focusing on the first step. Why? Because if it was done correctly, then all of the other steps followed naturally, like the flow of a river to the sea!

Like most other writers of instructions for solving the cube that can now be found on the internet, I knew that almost anyone could complete the first two steps by trial and error. But, aligning the second and third layers without destroying what you had already done, is much more difficult. However, now, after completing the cube many times, I realized that the first steps were actually the most important ones. If they were done correctly, then *all* of the other steps followed more easily, than when the first were done by trial and error. Also, I recalled that by modeling the basic units and structures of reality, mass, energy and gimmel, body, mind and soul, we saw how the cube reflects the patterns of atomic structure, the universe, and life. How could this new discovery of the importance of the first step relate to life and human consciousness?

We know that trying to solve the cube by trial and error without knowledge of effective algorithms could take forever; and similarly, trying to solve the problems of life without having propose, or any plan of attack, could take a lot longer than one lifetime. However, if we make a little progress in one lifetime, and the universe gives us as long as we need or want to take, by allowing us to return after a short rest, then after a while, we begin to remember more and more. Eventually, we know the minute we are born that we are not here to fool around and wind up in pain and sorrow, but that we need to do the first steps right, so we can get on with it. We realize that alignment, focus and perseverance are needed from the very start. This would explain why some people (and I was one), have memories of some of their previous existence and understand the importance of not wasting the energies of youth, or the mental clarity of early adulthood. If the first steps are not taken correctly, i.e., in accordance with the laws of nature, the body and mind will not last long enough for the alignment and focus needed to progress in later years. Could it be that each time we solve the cube, it is analogous to a single life?

One of the most basic laws of science, discovered by Russian scientist Mikhail Lomonosov in 1756, is the *conservation of mass*. Around 1850, James Clerk Maxwell intuited that there was a natural mass-energy equivalence. Other late-1800- early-1900 scientists, like Max Von Laue, agreed that this must be the case, and in 1905, Albert Einstein provided the theoretical physics and mathematics that led to the proof. With the discovery of gimmel as the third form of the essence of reality, the conservation of mass, energy *and consciousness* is a logical extension of the law. The mathematical logic of the Calculus of Dimensional Distinctions applied to the second law of thermodynamics and quantum physics in TDVP, reveals the fact that the very existence of the stable proton and the resulting universe proves that consciousness is primary. Therefore, I suggest that we must conclude that the essence of mass, energy *and consciousness* is conserved in all processes in the universe.

Conservation of Consciousness

There is no question that the essence of reality, whether mass, energy or the agent of consciousness in the form of gimmel, is conserved. And, because we have shown that consciousness is primary, it stands to reason that whatever progress toward enlightenment and Cosmic Consciousness you may have achieved in your lifetime is not lost. Therefore, the question, and it is *a very important one*, becomes:

How is the consciousness of the soul conserved, and in what form?

If you are not one of the estimated 1.4 billion people on the planet who believe in reincarnation, I am not asking you to suddenly believe in reincarnation. I'm asking you to look at the question with an open mind. Put whatever you may have been taught by your parents, your religion or mainstream science, and whatever you think you know, aside for now, and look at the evidence. Are human souls reincarnated? About 51% of the world's population believe in some form of survival of consciousness after the death of the physical body, and about 24% of Christians in the US believe in reincarnation. But belief and knowing are two different things. How can we prove or disprove the hypothesis of the survival of consciousness and/or reincarnation?

What does mainstream science have to say about it? For those scientists who have accepted the physicalist metaphysical belief system of materialism, the answer is that no form of survival is possible. But that is a belief, not science. It does not even rise to the level of a scientific hypothesis, as defined by scientists: A scientific hypothesis must be subject to proof or disproof. The belief that the universe could exist without consciousness cannot be proved because proof depends upon indisputable evidence of observation, measurement and logic, and no reality can be observed, measured or subjected to logical analysis without a conscious observer. Furthermore, the discovery of gimmel has proved that no reality could have evolved out of a big-bang explosion without the involvement of the non-physical gimmel as an organizing form of consciousness.

On the other hand, the hypothesis that consciousness survives the death of the physical body is a valid scientific hypothesis, and it can be proved. The evidence for it is far more logical, well-founded, legitimate and convincing than the evidence for the existence of the Higgs boson, or any particle of the particle zoo, other than the up-quark, down-quark, proton, and the electron! Dr. Ian Stevenson of the University of Virginia (who passed away in 2007) and others, have documented over 3,000 cases of children who have remembered past lives, not just in families in places like India, where virtually everyone believes in reincarnation, but also in families here in the US and other countries with no belief in reincarnation. The scientific evidence is actually overwhelming, but reincarnation, if it exists, cannot be the simplistic thing that people who try to disprove it, take it to be. The common arguments used against it here in the West are based in belief, not science, and they are easily debunked. I'll address some of the most prevalent ones here, but first, I must provide a little background.

The Misinterpretations of the Teachings of Jesus of Nazareth

Survival of the essence of the soul, or human consciousness is one of the simplest, yet most obscured and misunderstood ideas there is. One has to pursue an unbiased, open-minded study of the twisted history of the organized religious and political institutions of this world to understand why and how *ideas* regarding the survival of the soul have been distorted and obscured. The roots of Christianity, for example, are found in the spiritual practices of the Judeans over 2,000 years ago. Today's organized Christianity professes to embody the teachings of Jesus, who was born in a Jewish community in the area we call the Holy Land more than 2,000 years ago.

It is well-documented that the rebirth of souls was a basic belief of Judaism in the time of Jesus, and there is no evidence that Jesus discounted or rejected the idea, nor did any of his followers who lived during his lifetime. In fact, in documents quoting Jesus, we find him obliquely referring to the reincarnation of Elijah, Elisha and others, even in the highly redacted versions of the scriptures that we have today. Once you realize that the scriptures that we call the Bible today have been heavily edited by non-believers in powerful positions in the political and religious institutions of the past, and changed multiple times to suit their political and philosophical agendas, you can begin to see how some of the original teachings of Jesus may have been distorted and, in some cases, even deleted from the scriptures we have today.

The Roman Catholic Church claims that St. Peter was the first Pope of the Church, because Jesus named him as his successor, but it is very unlikely that Peter was ever actually part of the Roman Catholic Church, because Peter, like Jesus was a Jew. and he was crucified in Rome, **by Romans**, in 67AD, during the reign of the Emperor Nero. He was crucified in a horrific manner in the public square in Rome because he was identified as a Jewish leader of people following the anti-Roman teachings of Jesus of Nazareth. The teachings of Jesus challenged the pagan religion of the Roman state. Peter was replaced immediately, according to Church historical records, by Papa Linus, **a Roman**. It is obvious to any objective observer that this was a political move by the Emperor and the Roman government, to take control of the followers of Jesus, a rapidly growing branch of Judaism at the time.

There was no such entity as the Catholic church at the time of Peter's execution in 67AD. The term "Catholic" comes from the Greek word καθολικός, meaning universal, or "of the whole". It was coined by the Greek Christian theologian Origen around 200 AD. Predictably, from the time of Peter's execution, the popes of the Christian Church were either appointed by the Emperor or elected by the Romans, and until the end of Roman Empire, popes were always either Roman or Greek, never Jewish.

For several hundred years, Christian teachings included reincarnation, as they had from the beginning as a dissenting sect of Judaism. Origen, a Greek born in Alexandria, was the most prolific Christian writer of the third century AD, producing more than 6,000 treatises on Christian philosophy and theology, including commentaries on the Hebrew Bible and the teachings of Jesus recorded in scriptures that became known as the New Testament, as well as several other scriptures later rejected by the Catholic Church as Jewish or pagan. He wrote about reincarnation in two of his major treatises as follows:

"Each soul enters the world strengthened by the victories or weakened by the defects of its past lives. Its place in this world is determined by past virtues and shortcomings." - From Origen's work: *"De Principalis"*

"Is it not more in accordance with common sense that every soul for reasons unknown - I speak in accordance with the opinions of Pythagoras, Plato and Empedokles - enters the body influenced by its past deeds? The soul has a body

at its disposal for a certain period of time which, due to its changeable condition, eventually is no longer suitable for the soul, whereupon it changes that body for another." - From *"Contra Celsum"*

Origen was, and still is to this day, highly regarded as one of the most important Christian theologians of all time, and a founding father of the Catholic Church. So how did some of his teachings, especially the belief in the survival and continuation of souls get so vociferously eliminated from Church doctrine? The answer may surprise you: It was not a Pope or any member of the Catholic priesthood who decided to ban the doctrine of reincarnation from Church doctrine, it was Justinian, Emperor of the Roman Empire.

The Last "Great" Roman Emperor: Justinian I

By the year 500, the power and influence of the Roman Empire was beginning to fade. The Emperors of Rome, like the rulers of many civilizations before them, had gained their power by use of brute force and violence, and had maintained it by claiming that the line of emperors were direct descendants of the gods, i.e., in this case, Zeus and company. They used their wealth gained by killing, conquering and converting the peoples around them to slavery to perpetuate and glorify their gods and images of their emperors as descendants of the gods. But they were only human beings after all, and their absolute power gradually became absolute corruption.

The Emperor Justinian was a clever, well-educated and thoroughly evil man, also known as Justinian the Great, and would become known as Saint Justinian in the Eastern Greek Orthodox Church. His stated goal was to "revive the Roman

Empire's greatness and reconquer the lost western half of the historical Roman Empire". In his mind, the decline of Rome's influence in the western parts of the Roman Empire was due in large part to the ascending influence of Christian teachings, as disparate anti-Roman groups had coalesced under the teachings of Jesus of Nazareth, who was becoming a mythical legend, challenging the divinity of the line of Roman Emperors. Justinian was smart enough to see that he couldn't openly declare war on the offending churches, so he cleverly plotted to undermine their influence and integrate them into the Roman theocracy.

By the time of the Emperor Justinian, the decadence and debauchery of the rulers of the Roman Empire had been well-known for hundreds of years, and the ranks of the Christian Religious Sect had also been steadily growing during the same period of time. He carefully studied the writings of Origen, the most influential of the Christian theologians, and picked out a list of ideas put forth by Origen that he could use to subvert the teachings in the western provinces and integrate them into Roman theology.

Justinian's *Anathemas* Against Origen

Justinian realized that some of the teachings of the followers of Jesus constituted a serious threat to his power; e.g., according to Origen, Jesus had said: "Render unto Caesar the things that are Caesar's, and unto God the things that are God's". And "All are sons of the most-high God". And Origen had written: *"Each soul enters the world strengthened by the victories or weakened by the defects of its past lives. Its place in this world is determined by past virtues and shortcomings."* Such teachings were in direct conflict with what Justinian saw as his divine right to rule, so he seized on this statement and related ideas in Christian doctrine, as interpreted by Origen, that undermined the Roman ruler's claim of divinity. If people were led to believe that by being virtuous, they could rise to the level of an Emperor, i.e., to the status of a god, then the Emperor's power would be seriously threatened. He decided that he must declare this idea to be heresy and act to stamp it out. In what he called the ten *anathemas*, an edict that he prepared for this purpose, read in part:

"Whosoever teaches the doctrine of a supposed pre-birth existence of the soul, and speaks of a monstrous restoration of this, is cursed. Such heretics will be executed, their writings burned, and their property will become the property of the Emperor."

In 553AD, the Emperor called for an immediate assembly of a Council of the Church Fathers to ratify the decree, but the meeting of the Council was opposed by the Pope. The Emperor then cleverly forced several Eastern bishops to attend a secret meeting where he presented his '*Fifteen Anathemata*' to them, condemning Origen's writings. He prevailed upon them, under threat of death, to sign the decree. The meeting with the bishops prior to the Council was a bold ploy to undermine the Pope's power and promote the ban on the teachings of Origen. The scheme worked. An official meeting of the Council was held on the fifth of May 553, and the Pope was forced to accept the decree, allowing the Emperor to issue the ban as if it were imposed by the Pope and the Council of Bishops.

As a result, throughout Europe and the Middle East, monks educated as scribes were hurriedly put to work, urgently expunging all references to the pre-birth existence and transmigration of the soul from all of the existing versions of Biblical scripture, so that Christian dioceses throughout the land would not risk the wrath of Justinian, which they knew was a very real threat. The purge of references to rebirth was very thorough, and where it could not be eliminated without destroying whole passages vital to the teachings, it was re-worded to imply a spiritual rebirth, not a physical one.

In this way, the ban on the belief in reincarnation was forced into Church doctrine, and no attempt was made to rectify Justinian's self-serving actions until after the participants in the Council of 553 had passed away and even the memory of the fact that belief in reincarnation had once been part of Church doctrine had passed away. In this way, the ban was passed down to this day, and reincarnation has played no role in mainstream Christian thinking, in stark contrast with almost every other religion in the world, with the exception of the other Abrahamic religion that we know as Islam.

The account I've given here of the history of the banning of the concept of reincarnation from Christian thought comes from a number of reliable sources, including a Roman Catholic monk and member of a European royal family I met in 1960, who told me and a few others that he had seen the records of the purging of the mention of reincarnation from Church documents in the archives of the Vatican City in Rome. A German Biblical scholar wrote extensively about the expunging of reincarnation from Church doctrine. Here is a quote from the German Scholar:

"... the 'ban' on the teaching of reincarnation is based on historical misrepresentation and has no ecclesiastical authority. It was in fact a 'fait accompli', brought about by Justinian, which no-one within the Christian church has dared to challenge in the course of some 1500 years. What is worse is that the subject has been totally ignored, as a glance at any encyclopedia will show." (Peter Andreas: *Jenseits von Einstein*, Econ Verlag, Germany, April 1999)

Clearly, the Emperor Justinian was not a spiritual man, He was a political tyrant whose goals were to gain and keep power and wealth anyway he could. He enforced the ban on the teaching of reincarnation and initiated other heavy-handed rules of state on the leaders of the Church under threat of execution and confiscation of Church property. *His goal was to control people, not enlighten them.* The priests and the Pope acquiesced, and within a few decades, Justinian's oppressive rules became institutionalized as part of the dogma of the Catholic Church and the Church became one of the most powerful arms of the Holy Roman Empire, used to control the masses and most of Europe.

Justinian's clever scheme to pervert and integrate Christian doctrine with Roman paganism was successful far beyond his wildest dreams and personal goals, and reached far beyond his time. Justinian's *anathemata* were happily adopted and adapted by control-minded priests within the Roman Catholic Church and used to initiate and perpetuate the bloody *inquisitions* that swept across Europe in the 13th century AD. A group of institutions were established within the Catholic Church whose stated aim was to combat heresy. In fact, the institutions of inquisition were used to wage war against religious sectarianism that threatened the ruling power of the Holy Roman Catholic Church, in much the same way Justinian had used his power to re-conquer the parts of the Roman Empire that were converting to the original form of Christianity and integrate them into the Roman Empire in his time. Such diverse sectarian groups as the Cathars in France, the Hussites in Eastern Europe and the Waldensians in the Italian and Swiss Alps and Germany were targeted. Within the Catholic Church, the Spiritual Franciscans and other orders with whom the Dominicans, who dominated the inquisition institutions, disagreed, were also viciously targeted.

During the Middle Ages and the early Renaissance, the scope of the Inquisition was expanded in an effort to put down the Protestant Reformation and the Counter Reformation within the Catholic Church. This expansion resulted in the Spanish inquisitions, and consequently, the inquisitions throughout Spanish empires in Africa, Asia, and the Americas, including the Peruvian and Mexican inquisitions, where hundreds of thousands of natives were tortured and killed. Encouraged by their successes, the Spanish and Portuguese inquisitors focused on the Jewish and Muslim converts to Catholicism, under the assumption that they were still secretly practicing their former religions. This led to a further expansion of the inquisition as a means of eradicating whole communities of Jews and Muslims.

I am not trying to paint the Catholic Church as completely evil, as many of the leaders of the Protestant movements like Martin Luther, Jehan Cauvin (John Calvin) and Jan Hus did - and some protestants still do today. **Many priests and lay practitioners of the Catholic faith were and are spiritual people devoted to Christ and Christian principles**. And the story of the perversion of organized, institutionalized church and state into heavy-handed political entities, is by no means limited to the Catholic Church. It appears to be endemic in organizations that become dogmatic state-controlled institutions.

Such organizations seem to magnify human weaknesses and failures often bringing about unnecessary death and destruction. The Spanish inquisition of the 13th century, which gained the most attention with its many atrocities, was a long time ago, but the Institution of Inquisitions initiated by Justinian's integration of the Catholic Church and Roman Empire in the 6th century didn't end in the 13th, 14th, or even the 19th century. The Institution of Inquisitions still existed within the Catholic Church in the 19th century, and even survived the abolishment of the inquisition courts outside the Papal states in the late 1800s.

The Perpetuation of Evil from Justinian I to Adolph Hitler

The Institution of Inquisitions, like Justinian's anathemata, was created to stamp out heresy, i.e., anything that contradicted Church doctrine, and in spite of claims by Church spokesmen to the contrary, it has never gone away, it has been renamed as: "The Supreme Sacred Congregation of the Holy Office". Its purpose is essentially the same as it has always been: to suppress ideas that are counter to Church dogma, its stated objective is to "spread sound Catholic doctrine and defend those points of Christian tradition which seem in danger because of new and unacceptable doctrines." In the 1930s, Adolf Hitler became the latest and most powerful world leader to use the same evil logic of Justinian's anathemata to perpetuate violence and genocide.

Hitler was born to a Catholic mother in Braunau, an Austrian town near the German border, and was christened Adolphus Hitler in 1889. While he was a child, his family moved back and forth between Austria and Bavarian Germany. At the age of eight, he sang in the choir in the Catholic Church in the town where he lived, and after he was confirmed as a Catholic, he considered becoming a priest. But that was not to be his path in life. When the First World War broke out, he volunteered in the German Army. Injured in battle, he was awarded the Iron Cross and commended for bravery by his superiors. Hitler never accepted the idea that the German Army had been defeated, He blamed the capitulation on civilian leaders who had no stomach for war.

After the war, Adolf Hitler joined the National Socialist German Workers Party, where he became a confirmed socialist and anti-Semitic, and within which he eventually rose to power because of his tireless energy and impressive oratory skills. He took advantage of the bitterness of the German people over the heavy penalties imposed on them at the end of the war, blaming the British, the Russian Marxists and the Jews for Germany's military and economic collapse. For personal and political reasons, Hitler embraced the evil logic of the anathemata of the Emperor Justinian with great fervor as an excuse to continue the inquisitions started by the Catholic Church in the name of the Holy Roman Empire, and exact revenge on the Jews and other non-Aryan people in Germany who he blamed for all of Germany's woes. Hitler's vision was the establishment of the *Third Reich*, meant to be the new, emboldened version of Justinian's Rome and the Holy Roman Empire.

It might be argued that the same evil spirit that drove Emperor Justinian, may have reincarnated in Adolf Hitler. In 1933, when Mussolini, the socialist dictator of Italy, joined with Hitler's National Socialist Party to form the Third Reich Axis, he and Hitler signed an agreement called a *Concordat* with the Vatican in an effort to strengthen the power of the Nazi party with the Catholic Church as the spiritual arm of the new world government. This was a no-brainer for the Pope XI and Pope XII because it protected the Church's holdings in Germany and if Hitler succeeded in his quest to conquer all of Europe, which it looked as if he was likely to do, the socialist ideals of the Holy Roman Empire and the Catholic Church would be spread across the globe. If he didn't succeed, the Church could claim that the *Concordat* with Hitler was signed only to protect its churches in Germany and to avoid an invasion of the Vatican City, which Hitler certainly could easily have done. The Italian government, with the secret complicity of the Vatican, helped fund Hitler's war efforts, because it was a way to re-establish the Holy Roman Empire as an integral part of the new world government, the Third Reich.

Franz von Papen, a German Nobleman, Chancellor of Germany in 1932, a strong supporter of Hitler as he rose to power, and Vice-Chancellor under Hitler, was instrumental in setting up the *concordat* between Germany and the Vatican. Von Papen publicly announced that "The Third Reich is the first world power which not only acknowledges but also puts into practice the highest principles of the papacy." The Vatican considered the governments that had signed the concordat, i.e., Italy and Germany, as a part of the "Government of God", and the Vatican swore to stabilize and uphold such governments and give them divine protection.

Hitler wrote this about his Third Reich:

"I learned much from the Order of the Jesuits. Until now, there has never been anything more grandiose, on the earth, than the hierarchical organization of the Catholic Church. I transferred much of this organization into my own party."

After the defeat of the Axis in the Second World War, the connection of Hitler's Nazi party, the "Third Reich" with the Catholic Church was largely covered up for obvious reasons. It is interesting to note that, because Hitler became such an icon of evil in the western world, the word Nazi, pronounced "nahtzee" in German, has become a popular symbol for pure evil. In fact, NAZI is simply an abbreviated contraction formed by taking prominent letters from "**Na**tionalso**zi**alistische", the key word in the official name of the German national socialist party in German, which is: "Der Nationalsozialistische Deutsche Arbeiterpartei", which translates in English to "The National Socialist German Workers Party".

Most people certainly do not think of Adolph Hitler as a Christian, but he was a life-long member of the Catholic Church. The following, from Hitler's reassurance to the Pope Pius XII about the safety of Christian churches under his Reich provides a glimpse of how he related to Christianity personally:

"I am personally convinced of the great power and deep significance of Christianity, and I won't allow any other religion to be promoted. That is why I have turned away from Ludendorff and that is why I reject that book by Rosenberg. It was written by a Protestant. It is not a Party book. It was not written by him as a Party man. The Protestants can be left to argue with him ... As a Catholic I never feel comfortable in the Evangelical Church or its structures. That is why I will have great difficulty if I try to regulate affairs of the Protestant churches. The evangelical people or the Protestants will in any case reject me. But you can be sure: I will protect the rights and freedoms of the churches and not let them be touched, so that you need have no fears about the future of the Church."

At the end of the war, when the world press announced that Hitler was dead, Franco, the military dictator of Spain who had also signed the concordat with the Vatican in his effort to establish a Socialist Catholic Monarchy, published the following:

"Adolf Hitler, son of the Catholic Church, died while defending Christianity. ... Over his mortal remains stands his victorious moral figure. With the palm of the martyr, God gives Hitler the laurels of Victory."

Uncovering the Truth

Going back to the original Hebrew, Aramaic and Greek writings from which the first Biblical scriptures were carefully selected for political reasons, we can study the scriptures of the Bible in context with the scriptures that were eliminated by the Roman Empire and the Catholic Church for political purposes. And, if we study the writings of Origen, the first comprehensive analysis of early Christian teachings, then we can come much closer to having a true understanding of the teachings of Jesus in their proper historical and cultural context.

Origen was by far the most prolific Christian writer in the first 200 years after the time of Jesus, and his works were not completely destroyed in spite of Justinian's diabolically clever efforts. His writings fill in the gaps in the real Christian history from the time of Jesus until the organized institutions of state and church colluded and conspired to twist his teachings into dogma designed to support their political agendas and control the masses with the goal of perpetuating their power and wealth. Let's look at what we know about Origen.

Origen, Outstanding Biblical Scholar

Origen lived from 185 to 254AD, and he spent the first half of his life in Alexandria, where he contributed many scholarly treatises to the store of knowledge in the famous library there. He is said to have been "the greatest genius the early church ever produced" (*John Anthony McGuckin (2004). The Westminster Handbook to Origen. Westminster John Knox Press. p. 64. ISBN 978-0664-22472-1*, page 25.) In Alexandria, Origen was under the direct supervision and orders of Demetrius, the Bishop of Alexandria, who, unfortunately, was a willing puppet of Rome.

According to the Roman historian Herodian, the brothers, Caracalla and Geta, who vied for the position of Emperor when their father, Emperor Severus, died in 211, were as different as night and day: Caracalla, the favorite of Severus, was a brutal military leader who would slaughter a few thousand civilians at the drop of a hat, while Geta, their mother's favorite, was of a more compassionate and considerate nature. True to Herodian's appraisal of the two, Caracalla murdered his brother Geta to become emperor.

In 215, after brutal and bloody military campaigns in Britain and Germany, Caracalla visited Alexandria. During the visit, a group of intellectuals and students confronted him, protesting the fact that he had murdered his brother Geta. Caracalla was incensed and ordered his troops to ravage the city, execute the governor, and kill all the protesters. He also commanded them to expel all of the teachers and intellectuals from the city. Origen fled to Caesarea, where his fame as a scholar and natural philosopher had preceded him. The elders and citizens of Caesarea welcomed him with open arms and eagerly attended his lectures on Christian principles.

Demetrius ordered Origen back to Alexandria, and issued a decree condemning the Caesareans for allowing Origen to preach without his permission. Origen complied with Demetrius' order and returned to Alexandria, bringing with him an ancient scroll of the Torah, said to have been found sealed in an amphora, inscribed in the original Hebrew, which he studied at great length and translated into Greek. He had also studied the books of the New Testament more thoroughly than any other scholar had at that time. He was arguably the most learned scholar of the era, and without doubt the most knowledgeable regarding the teachings of Jesus.

Origen repeatedly asked Demetrius to ordain him as a priest, but Demetrius refused, ostensibly because he knew Origen possessed a deeper understanding of Christian philosophy than he did. In 231, Demetrius sent Origen on a mission to Athens. On the way, Origen stopped in Caesarea to visit the Christians he had met there when he fled Alexandria in 215. He asked the Caesarean Bishop Theoctistus to ordain him as a priest, and Theoctistus, who admired Origen's intellect and scholarly writings greatly, was more than happy to ordain him.

Upon hearing about this, Demetrius was outraged, and issued a condemnation declaring that Origen's ordination by a foreign bishop was an unconscionable act of insubordination. However, being ordained as a Catholic priest, Origen no longer had to answer to Demetrius, so he stayed in Caesarea and built a library there and founded the Christian School of Caesarea, where he taught logic, cosmology, natural history, and theology. He was quickly regarded by the churches of the area along the Jordan River, the Dead Sea and northern Arabia as the ultimate authority on all matters of theology.

In 250, the Emperor Decius, declaring that a plague afflicting the area at that time was caused by Christians not recognizing his divinity, issued an edict that everyone in the Roman Empire had to perform a sacrifice to the Roman gods for the well-being of the Emperor. Origen refused and was arrested and tortured. He was flogged, his neck pierced with the inner spikes of an iron collar and he was stretched on the rack for days on end. The Roman Governor gave strict orders that Origen was not to be killed until he had publicly renounced his faith in Jesus as the Christ, and all Christian teachings. He refused, declaring that he would not renounce his faith, even under pain of death. Eventually, to avoid the Governor's orders not to kill him until he recanted, the tormentors stopped torturing him. But, within a short time, in 253 or 254 AD, he died from the injuries inflicted during the tortures.

Roman Destruction of Libraries

The City of Alexandria was attacked, and the library was burned by order of Roman Emperor Aurelian in 273AD, and the Christian library that had been established by Origen in Caesarea, that had been protected and expanded by the scholarly presbyter Pamphilus, became almost as famed as the library in Alexandria. In 630 it was said to hold more than 30,000 manuscripts, including thousands of Origen's works, largely originals written in Origen's own hand. Unfortunately, the Caesarea Library was also burned by the Romans sometime around 635AD, and the destruction of any Judeo-Christian manuscripts left behind by the Romans was completed by invading Saracens, who captured Caesarea in 638AD.

The library at Antioch in ancient Syria was burned by Emperor Jovian in 363AD because it contained manuscripts considered by Jovian to be pagan.

The Holy Roman Empire was certainly not the only political system of government to distort and destroy valuable knowledge. History shows that virtually every civilization on the planet has done the same thing. This practice of elevating political institutions and forcing organized religious institutions to serve brutal political rulers, has perverted the teachings of enlightened beings, obscured the real purpose of life and spirituality, and played into the materialistic belief system dominating governments and mainstream science today.

All, however, is not lost. In fact, the universe (Primary Consciousness) is always in control, and puts the final responsibility for spiritual, intellectual and social advancement on the individual, which is where, after all, it should be.

Spirituality and compassion cannot be legislated and enforced upon unwilling subjects, or decreed from a pulpit. It has to start in the human heart and expand to the mind to find proper physical expression. Any other route is artificial and doomed to failure. The original teachings of the most enlightened spiritual thinkers of the world include the pre-existence of souls and their continued survival until they gain enlightenment by proper effort and alignment with universal law. Organized religion, politics and now the belief in scientific materialism continue to try to suppress the human spirit and obscure the existence of Primary Consciousness.

Arguments Against Reincarnation

With this brief historical background, let's look at the most common arguments put forth against the reincarnation hypothesis by some Christians, Jews, Muslims, Agnostics, Atheists, and Materialists today. Interestingly, the belief that reincarnation cannot be possible is one of the few things that some members of all of these groups might agree on - but for quite different reasons. Before we get into the arguments against reincarnation, some general observations are in order.

First, arguing that reincarnation does not or cannot occur, is an attempt to prove a negative. It certainly is not impossible to prove a negative, but it is generally much more difficult than proving a positive proposition. It is also a difficult and even dangerous position to defend logically because you may have dozens of reasons why you *believe* reincarnation does not happen, and you may be able to spend hours explaining each one of them very articulately, but it only takes one indisputable example to prove you wrong. As William James famously said: "If you wish to upset the law that all crows are black, you mustn't seek to show that no crows are white; it is enough if you prove one single crow to be white."

Second, there are Christians, Jews and Atheists who do believe in reincarnation, but most devout Muslims will tell you that a Muslim cannot believe in multiple reincarnations, because the Koran only allows one, and that, according to the Prophet Mohamed, happens on Judgement Day. However, some Muslim holy men who claim to have attained enlightenment, teach that reincarnation does happen. One such holy man explained to me that Judgement Day is not just one day for all souls, but occurs for each soul at the time of each death. When I asked him about the verses in the Koran that seem to dispute that, he said: "The words of the Prophet Mohamed, *may He rest in peace*, are for the common man, not enlightened saints".

Third, Atheism and belief in reincarnation are not mutually exclusive. An Atheist can believe in reincarnation as a natural process that may occur without requiring the existence of a god. So even within groups that generally do not believe in reincarnation, there are people who do believe in reincarnation. With that, let's turn to the arguments against reincarnation. There are plenty to be found in books and articles, and on the internet.

Faith-Based Arguments

Faith-based arguments are usually not scientific arguments, because they do not begin by considering reincarnation as a hypothesis to be proved or disproved. They start with the assumption of superior knowledge, by basing their argument on specific scriptures, such as verses of the Torah, Koran or Bible, which they consider to be the Word of God. Unfortunately, arguments presented in this manner are circular because they have already assumed the negative conclusion they seek to prove. Such arguments are simply arguments in defense of a point of belief or faith in a religious doctrine, and therefore can be accepted or rejected, depending on whether or not you share a belief in the dogma of the presenter's faith.

It is not my intent to belittle anyone's faith, or to dismiss their arguments against reincarnation because of it, but such proofs must be considered in their proper context. The person who presents faith-based arguments is relying on what he/she believes to be unimpeachable authority. Because I am writing in American English, and because I live in a place and time where Christianity is the prevalent faith, I will first put the Bible, the written authority upon which Christians rely, into its historical context. In other words, let's look at how the Bible, in particular the King James Version (KJV) and its many modern re-interpretations, came to be what they are today.

The Origin and Evolution of English Versions of the Bible

Arguments depending upon the belief that specific scriptures are the infallible word of God should be viewed with some skepticism, because, as we've seen, the elimination of the teachings of Origen, arguably the brightest of the early church theologians, by the Roman Emperor Justinian's ban, has changed the content of the Christian Bible substantially. And, if we accept the claim that the scriptures upon which the New Testament is based were originally the word of God, spoken by Jesus Christ, a truly Enlightened Being, then we still must recognize that it has come down the ages through the lenses of many less than perfect human beings. For example, the King James Version of the Bible, revered by the fundamental Protestants of the hill country where I grew up, as the infallible word of God, was authorized by King James as the head of the Church of England in 1604. But the King James Version was not the first translation of the Bible into English from the original Hebrew, Aramaic and Greek. It was, in fact, the fourth.

The first English translation of the Bible, called "The Great Bible", was commissioned by King Henry VIII, hardly a model of Christian virtue. After his first wife, Catherine proved unable to bear him a son, Henry requested that the Pope allow him to divorce Catherine and marry his mistress. The official position of the Catholic Church was that divorce was a sin, so the Pope refused. In response, Henry renounced the Catholic Church and married his mistress. He then proceeded to close all of the Catholic monasteries in England, seize the Church's assets and establish the Church of England with himself as its head, and in 1536, as an act of defiance, literally thumbing his nose at the Pope, he authorized the translation of the Bible into English as the official Word of God, an action strictly forbidden by Rome. Henry continued to do whatever he pleased, producing children by various wives and mistresses, trying to obtain a male heir, and he imprisoned and tortured anyone who opposed his actions, executing many, even including two of his six wives.

The second English translation of the Bible, now almost forgotten, was the Geneva Bible, produced in 1557 to1560. This version of the Bible came about when Henry the Eighth's only legitimate son, Edward VI, died after only six years on the Throne, and his older sister Mary became Queen of England and Ireland from 1553 to 1558. Mary was Catholic, and in order to reinstate the Catholic Church in England, she heavily persecuted and executed many English Protestants, earning the title "Bloody Mary". More than 800 English scholars fled to Europe to escape her Catholic wrath. They gathered in Geneva Switzerland in 1557 and proceeded to produce a new Protestant version of the Bible.

The Geneva Bible reflected the thinking of a movement of the time known as Calvinism, just one among several emerging protestant sects, including the Lutherans, Presbyterians and Episcopalians. The Geneva Bible was considered to be a threat to all Christianity by the bishops of the Church of England because it replaced the government of the church by monarch appointed bishops, with government by lay elders. After Bloody Mary's death, her half-sister Elizabeth became Queen. She was a Protestant and as Queen, the head of the Church of England.

With her blessings, the bishops of the Church of England, denounced the Roman Catholic Bible and the Geneva Bible as heretical and produced their own version, which became known as the Bishops' Bible. Produced under the authority of the Church of England in 1568, the Bishop's Bible succeeded Henry the Eighth's "Great Bible" as the official Bible of the Church of England. The Bishop's Bible was substantially revised in 1572, and with minor changes in the spelling of some Hebrew names in 1602, it was used as the base text for the King James Version (KJV) of the Christian Bible completed in 1611.

If this were anything but the Bible, no one would imagine that the KJV could possibly be the unaltered word of God as given to the Jews in the Torah and the unaltered Gospel as spoken by Jesus in the form of the New Testament. The KJV is a version of original Christian and pre-Christian scriptures, filtered through several secular interpretations, and all of them that occurred after the scholarly interpretations of Origen, who from all reports was a deeply spiritual man, were for political purposes and ego-based agendas whose instigators were far from virtuous.

The KJV's twisted past was completely unknown to most of the fundamental Christians in the USA of my childhood and teen-age years (1936 – 1955), and probably to most of the Protestants around the world. I remember statements of some of the good people of the hill country where I grew up when told about the history of the translations of the Bible, were asked how they could say that the KJV Bible was the original word of God. They said something like: "Yes, Priests, Kings and scholars may be less than perfect, but the changes and interpretations were guided by God and God would not allow his word to be distorted. The KJV is the only true word of God."

Confronted with this argument, I thought: Why would God choose power-hungry politicians, murders, ego-maniacs, atheists, adulterers, and sinners of every sort to shape His word, instead of honest scholars and spiritual people? But I didn't say it, because I knew what their answer would probably be: "God works in mysterious ways!" There just is no arguing with that kind of logic!

The Koran, the Holy Book of Islam

A brief look into the history of the organized Roman Catholic and Protestant Bibles has shown us that any argument for or against reincarnation based upon specific wordings in the Latin or English versions of the Bible are questionable, and the same can be said for arguments against reincarnation by Muslims based on verses of the Koran. The root word of Koran (in Arabic القران) is either 'Q'ar', meaning to collect, or 'Q'ara', meaning to recite. Both of these roots seem to fit the facts of the origin of the Koran because the Islamic Sacred Scriptures were not written down by Mohamed, but recited from memory. Several years after the Prophet died, his followers began jotting their memories of the recitations they had heard on camel bones and scraps of paper, and at some point, they were collected and hand-copied and bound into the form of a book.

According to Muslim belief, the revelation of the Koran (also spelled 'Quran' in English to approximate the guttural sound of the Arabic consonant) began in 610 AD, when the angel Gabriel (Arabic: جبريل, *Jibrīl* or جبرائيل, *Jibrāʾīl*) appeared to Muhammad in Hira Cave near Mecca, reciting to him the first verses of *Sura Iqra* (*al-`Alaq*). Throughout his life, Muhammad continued to recite revelations until his death in 632. The Quran as it exists today was compiled into a book format by Zayd ibn-Thabit and other scribes under Uthman, the third Caliph, a political leader of the Islamic Caliphate, a theocratic government, sometime between 644 and 656. This was about 100 years after the reign of Emperor Justinian of Rome, and about 450 years after Origen of Alexandria translated the original scriptures of the Torah and the collected sayings of Jesus and wrote his interpretations of the scriptures, much of which became the basis of the Christian Bible as it began its torturous transformation into the Bible we have today.

Faith-based arguments are not scientifically valid arguments. They can be accepted as true *only* if the tenants of the faith in question are accepted, either on the basis of personal experience, or by accepting the authenticity of someone else's experience and the authority of an institution established by them. So, as a scientist, I must remain skeptical of such arguments. This doesn't mean that I reject the teachings of Jesus. I certainly do not. On the contrary, I believe Christ Consciousness is real, and that it is the only road to Cosmic Consciousness. But faith is not a basis for proof, believing is not knowing, and science must prove the reality of something before announcing it as truth. A scientist cannot accept a concept as indisputable truth based on someone else's experience or belief. A scientist must have direct proof. Without direct proof, an idea, however appealing, is just a hypothesis; a theory to be tested, nothing more.

Unless you have direct two-way communication with Christ, you have no proof and must rely on hear-say and very questionable authority, because Jesus, like Mohamed, wrote nothing down. Neither did Gautama Buddha. Isn't this remarkable?

Is it possible that the founders of three major world religions were illiterate? No. They wrote nothing down because they knew that words in any language, misrepresent and distort as much as they reveal. Their revelations of the truth were much more complete, and on a much deeper level of consciousness than can ever be conveyed in words. It was their inspiring presence and extraordinary spiritual energy that convinced their followers that they knew the Truth, as much or more than the words they spoke. So, let's leave faith-based "proofs" behind and move on to non-sectarian arguments against the idea of reincarnation. Can science prove or disprove the reality of reincarnation?

Non-Faith-Based Arguments

The most convincing non-sectarian arguments against reincarnation come from those scientists who are materialists, atheists or agnostics. These groups are not mutually exclusive, but they are not synonymous either. A materialist can believe in God as a higher intelligence emerging from an evolving physical universe, and an agnostic, by definition, accepts the possibility of the existence of a higher intelligence, but remains skeptical until he or she sees proof. Atheism, on the other hand, is the completely negative position that there never was a god, is no god, and never can be a god. Obviously, this is a belief, not a scientific hypothesis, because it cannot be proved or disproved.

Rational arguments put forth by materialistic, agnostic, and atheistic scientists boil down to two positive statements:

1. Reincarnation produces an unreconcilable paradox of numbers
2. There is no credible evidence

Is There a Paradox of Numbers?

The fact that there are many, many more people alive on the planet today than at any time in the recorded past, is given by some skeptics as an argument against reincarnation, and at first glance it may seem like a good argument. However, on closer examination, it does not eliminate reincarnation as a logical possibility, because even if there were only a finite number of souls, say 10 billion, the assumption that all of them would eventually be on the planet at the same time is unwarranted. Also, many more people have died during the recorded past than exist on Earth today, so it is logically possible that everyone alive today may have lived before. So, there is no paradox of numbers.

Science is by definition, a search for truth. To determine whether an idea is true, false, meaningless, or beyond our ability to determine, a scientist must first frame it in the form of a hypothesis that may be falsified, like William James' statement "all crows are black". We can do that with the question of whether reincarnation is a reality or just wishful thinking with the following *hypothesis*: **The consciousness of an individual sentient being is wholly produced by that individual's physical body and brain, and cannot exist without them**. If this hypothesis is true, then when the body and brain of an individual cease to function, or when they are destroyed, by whatever means, the consciousness of that individual is simply gone. It can no longer exist, period.

As professor James pointed out, there is no need to look at all of the arguments that may be made supporting this hypothesis. If there is even one counter example, the hypothesis is invalid.

Evidence for Reincarnation

The strongest rational argument for reincarnation is based on extending the logic of the laws of cause and effect and conservation of substance, which apply to mass and energy, to include consciousness. The discovery of the existence of the impact of consciousness as gimmel in every stable structure in the universe establishes the link between mass, energy and consciousness, suggesting that consciousness is subject to conservation and cause and effect. But the final establishment of the reality of reincarnation depends on the documentation of indisputable evidence.

Scientific Evidence

The largest body of scientific evidence of the transmigration of souls is found in the life's work of Dr. Ian Stevenson (1918-2007). Dr. Stevenson was a professor and research psychiatrist at the University of Virginia School of Medicine for 50 years. He was Chair of the Department of Psychiatry from 1957 to 1967, the Carlson Professor of Psychiatry from 1967 to 2001, and a Research Professor of Psychiatry from 2002 until his death. He was also the founder and Director of the University of Virginia's Division of Perceptual Studies.

Dr. Stevenson is internationally recognized for discovering and documenting evidence that memories and physical injuries can be transferred from one lifetime to another. He traveled extensively over a period of 40 years, investigating approximately 3,000 cases of children around the world who recalled having past lives. His meticulous research revealed evidence that children who recalled past lives also had unusual abilities, illnesses, phobias and familiarities which could not be explained by the experiences and environments of their current lives or heredity. The following is a summarization of one of the cases investigated and documented by Dr. Stevenson.

The Case of Swarnlata Mishra

Swarnlata Mishra started talking about memories of a previous life when she was 3 years old. Her memories contained many details that enabled Dr. Stevenson to locate the family of the deceased person she said she remembered being, and in the course of the investigation, she remembered more than 50 specific facts that were verified. Her case was different than many of Dr. Stevenson's investigations in that her memories were happy memories rather than memories of violent and traumatic events and they did not fade away as she grew older.

Swarnlata Mishra was born in Pradesh India in 1948. When she was just three years old, she told her father about her previous life in the town of Katni more than 100 miles from their home. She related many details of her previous life in Katni. She said her name was Biya Pathak, and that she had two sons. She recalled details of their home in Katni, including the following: "it was white with black doors fitted with iron bars; four rooms were stuccoed, but other parts were less finished; the front floor was of stone slabs." The house was in the Zhurkuti District of Katni; behind the house was a girl's school, in front was a railway line, and lime furnaces were visible from the house. She added that the family had a motor car (a very rare item in India in the 1950's, and especially before Swarnlata was born). Swarnlata said Biya died of a "pain in her throat" and was treated by Dr. S. C. Bhabrat in Jabalpur. All these details were written down when Swarnlata was three, and they were verified later when Swarnlata was 10 years old and they actually traveled to Katni. Until this, the two families were unaware of each other's existence. After learning of Swarnlata's claims, In the summer of 1959, Biya's husband, son and eldest brother traveled to the town where the Mishras lived with the intention of testing her to see if she really was a reincarnation of their Biya. They enlisted nine strangers to accompany them to the Mishras' home to pose as friends or family members that Biya had known well. Ten-year-old Swarnlata quickly picked the real family members from among the imposters and stopped in front of Biya's husband, lowering her eyes as Indian wives do in the presence of their husbands. Many other factual verifications in this case, hard to explain by any theory other than reincarnation, are found in the case files and in Dr. Stevenson's book, "Twenty Cases Suggestive of Reincarnation", University Press of Virginia. Also see "Children Who Remember Previous Lives" written for the layperson, McFarland & Company, Inc., Publishers, 2001.

Could this case be a "white crow" disproving the hypothesis that: **The consciousness of an individual sentient being is wholly produced by that individual's physical body and brain**. If you accept the work of Dr. Stevenson, a scientist whose work in other areas is not questioned, with the same level of skepticism exhibited by a particle physicist reviewing the evidence for the existence of the Higgs boson, then you would have to say that it is. Critics of Dr. Stevenson's reincarnation investigations like the phrase "extraordinary claims require extraordinary evidence" made popular by Carl Sagan; but just how extraordinary does the evidence have to be? I submit that for most of the critics of the study of reincarnation, no amount of evidence will ever be enough, because their objections are belief-based, derived either from a religious faith, or from a belief in simplistic scientific materialism. Dr. Stevenson documented some 3,000 cases, most of which are very difficult to explain any other way.

Why is reincarnation considered to be an extraordinary claim? How can something that is a fundamental belief of 1.4 billion people (Hindus, Buddhists, Jains, Taoists, Sikhs, and Shinto followers) be "extraordinary"? It is considered extraordinary mainly in western cultures, but according to data released by the Pew Forum on Religion and Public Life from a 2009 survey, 24 percent of American Christians believe in reincarnation.

Credible Evidence from Adults

Dr. Stevenson's files on reincarnation primarily contain cases of children who report having memories of past lives. The reason Dr. Stevenson focused on children is easy to understand. As a scientist born, raised and trained in a society and scientific community that largely rejected the idea of reincarnation on the basis of religious dogma and/or materialism, in order to undertake a serious Investigation of the reincarnation hypothesis, he had to behave as a true skeptic and allow the possibility that the hypothesis might be in fact, either true or false. If true, the most likely place to find evidence would be in the newly born. As a medical doctor and psychiatrist, he knew that the clarity of the memory of an event generally fades with the passage of time; and therefore, if it is possible that some people live more than one life, and if it is possible that some memory of past life events stored in the brain of the deceased can carry over into the brain of the newly born body, then it is most likely to surface shortly after birth, and to be expressed by the child as soon as he or she begins to talk.

If such memories occur, they are likely to fade with time, and become categorized as dreams by the individual, as the body and mind go through the overwhelming stages of growth, including the emotions of puberty, and the influences of other people. Also, an adult may suppress or exaggerate such memories, depending on his societal conditioning and beliefs. As a scientist, Dr. Stevenson was breaking new ground for western science, so he could not allow preconceived beliefs or opinions about why, or how reincarnation might occur, affect the investigations. This explains why Dr. Stevenson focused primarily on reports of other-life memories by children. But, are there adults who claim to remember past lives?

Some Famous People Who have Professed Belief in Reincarnation

The list of 25 famous people below is only a partial list, consisting of quotes that are readily available from public statements and published writings. There are many more who believe in reincarnation as completely logical, or from direct experiences remembered in this life.

Benjamin Franklin

"When I see nothing annihilated (in the works of God) and not a drop of water wasted, I cannot suspect the annihilation of souls, or believe that He will suffer the daily waste of millions of minds ready-made that now exist, and put Himself to the continual trouble of making new ones. Thus, finding myself to exist in the world, I believe I shall, in some shape or other, always exist; and, with all the inconveniences human life is liable to, I shall not object to a new edition of mine, hoping, however, that the errata of the last may be corrected."

Mark Twain

From his autobiography: "I have been born more times than anybody except Krishna."

Albert Schweitzer

"Reincarnation contains a most comforting explanation of reality by means of which Indian thought surmounts difficulties which baffle the thinkers of Europe."

Henry Ford

"I adopted the theory of Reincarnation when I was twenty-six. Religion offered nothing to the point. Even work could not give me complete satisfaction. Work is futile if we cannot utilize the experience we collect in one life in the next. When I discovered Reincarnation, it was as if I had found a universal plan, I realized that there was a chance to work out my ideas. Time was no longer limited. I was no longer a slave to the hands of the clock. Genius is experience. Some seem to think that it is a gift or talent, but it is the fruit of long experience in many lives. Some are older souls than others, and so they know more. The discovery of Reincarnation put my mind at ease. If you preserve a record of this conversation, write it so that it puts men's minds at ease.

I would like to communicate to others the calmness that the long view of life gives to us." - The San Francisco Examiner, 1928

William James

Renowned American psychologist and philosopher, William James delivered a significant science-based lecture, called "Human Immortality", at Harvard, in 1893. He later expanded his concepts to specifically include reincarnation. On this he wrote:

"... I am the same personal being who in old times upon the earth had those experiences."

Ralph Waldo Emerson

"The soul comes from without into the human body, as into a temporary abode, and it goes out of it anew it passes into other habitations, for the soul is immortal. It is the secret of the world that all things subsist and do not die, but only retire a little from site and afterward return again... Jesus is not dead; he is very well alive; nor John, nor Paul, nor Mahomet, nor Aristotle; at times we believe we have seen them all, and could easily tell the names under which they go."

Henry David Thoreau

Thoreau wrote in "Letters":
"I lived in Judea eighteen hundred years ago, but I never knew that there was such a one as Christ among my contemporaries."

Voltaire

"It is not more surprising to be born twice than once; everything in nature is resurrection."

Walt Whitman
In "Song of Myself", the famous poet wrote:
"And as to you, Life, I reckon you are the leaving of many deaths, (No doubt I have died myself ten thousand times before.)"

Thomas H. Huxley wrote in "Essays Upon Some Controverted Questions":
"I am certain that I have been here as I am now a thousand times before, and I hope to return a thousand times."

General George S. Patton
American World War II general spoke of memories of a number of past lives There are numerous reports of General Patton talking about being reincarnated. He believed that he had always been a warrior in one form or another. During World War I, he told his mother that he had been reincarnated. Later in life, he said: "So as through a glass and darkly, the age long strife I see, Where I fought in many guises, many names, but always me."

Napoleon Bonaparte
The "Little" Emperor believed that he had been born many times. He is reported to have discussed who he had been in previous lives with many people. Napoleon died in 1821. Twenty-eight years later, Adolf Hitler was born. Both men tried to take over Europe using the same methods, fought Russia to a loss, and were defeated in nearly the same way. Both were also considered to be the anti-Christ during and after their lives by many people. Could it be that Hitler was his next incarnation?

Paramahansa Yogananda

The founder of Self-Realization Fellowship wrote in his Autobiography of a Yogi":

"I find my earliest memories covering the anachronistic features of a previous incarnation. Clear recollections came to me of a distant life in which I had been a yogi amid the Himalayan snows. These glimpses of the past, by some dimensionless link, also afforded me a glimpse of the future."

Salvador Dali

The famous Spanish artist remembered several of his previous lives. He spoke of being St. John of the Cross in a previous life: "as for me, ... I am also the reincarnation of one of the greatest of all Spanish mystics, Saint John of the Cross. I can vividly remember my life as Saint John . . . of experiencing divine union, of undergoing the dark night of the soul . . . I can remember many of Saint John's fellow monks.

Shirley McLaine

On her website, Shirley says: "When I walked across Spain on the pilgrimage called the Santiago de Compostela Camino, I encountered myself in a former life. I discovered a part of me that lead to a greater understanding of myself. I also realized the karmic importance of some of the people that have been close to me in this existence. These realizations, and numerous others, have helped, inspired and added to my whole being. They have assisted in my better understanding myself and those around me. It doesn't matter if this type of realization is imagination or if it is memory. It is a truth that I have experienced on some level, in some form of reality and I embrace it as a gift from the Divine.

Three quarters of the Earth's people believe they have lived before and will live again; thereby enabling their Soul's journey a continuous learning experience. Stories abound regarding how people find each other again - for good or otherwise."

Sylvester Stallone

Sly Stallone is sure he had at least four past lives, and he experienced a gruesome end in one of them. In an interview early in his career, he said, "I'm quite sure I lost my head in the French Revolution." His success with his screen persona Rocky Balboa may have something to do with Stallone's claim that he was actually once a boxer who was killed by a knockout punch in the 1930s.

John Lennon

"I'm not afraid of death because I don't believe in it. It's just getting out of one car, and into another."

George Harrison

"Friends are all souls that we've known in other lives. We're drawn to each other. Even if I have only known them a day., it doesn't matter. I'm not going to wait till I have known them for two years, because anyway, we must have met somewhere before, you know."

Edgar Cayce

According to those who knew the 'Sleeping Prophet' and studied his readings, "Edgar Cayce found that the concept of reincarnation was not incompatible with any religion, and actually merged perfectly with his own beliefs of what it meant to be a Christian. Eventually the subject of reincarnation was examined in extensive detail in over 1,900 Life Readings."

Carl Jung

"This concept of rebirth necessarily implies the continuity of personality. Here the human personality is regarded as continuous and accessible to memory, so that, when one is incarnated or born, one is able, at least potentially, to remember that one has lived through previous existences, and that these existences were one's own, i.e., that they had the same ego form as the present life. As a rule, reincarnation means rebirth in a human body.

What happens after death is so unspeakably glorious that our imagination and our feelings do not suffice to form even an appropriate conception of it… The dissolution of our time-bound form in eternity brings no loss of meaning."

Jack London

London, author, best known for book "Call of the Wild", wrote: "I did not begin when I was born, nor when I was conceived. I have been growing, developing, through incalculable myriads of millenniums. All my previous selves have their voices, echoes, promptings in me. Oh, incalculable times again shall I be born."

Arthur Schopenhauer

"Were an Asiatic to ask me for a definition of Europe, I should be forced to answer him: It is that part of the world which is haunted by the *incredible delusion* that man was created out of nothing, and that his present birth is his first entrance into life."

Cicero

A Roman Nobleman (106 B.C. - 43 B.C.) who is considered one of the great philosophers of that time. In his composition, "On Old Age", he wrote:

"The soul is of heavenly origin, forced down from its home in the highest, and, so to speak, buried in earth, a place quite opposed to its divine nature and its immortality... It is again a strong proof of men knowing most things before birth, that when mere children they grasp innumerable facts with such speed as to show that they are not then taking them in for the first time, but remembering or recalling them."

Josephus

(Well-known Jewish historian from the time of Jesus)
"All pure and holy spirits live on in heavenly places, and in course of time they are again sent down to inhabit righteous bodies."

Jesus of Narareth

Perhaps most shocking of all for Christians today, is the fact that Jesus talked about reincarnation. We know from the writings of Origen and others that he spoke to his followers about reincarnation. But these records were banned from the Bible by the Roman Emperor and others for philosophical and political reasons. Under the edict of Justinian, monastic scribes expunged overt references to reincarnation from the scriptures, but some reincarnation references by Jesus that could be explained as *special circumstances,* were left in. For example, from Luke 9:18 – 21. [My comments are *italicized* in brackets]

And it came to pass, as he was alone praying, his disciples were with him: and he asked them, saying, whom say the people that I am?

They, answering said, John the Baptist; but some say, Elijah; and others say, that one of the old prophets is risen again.

[*This is a clear reflection of the fact that the Jewish people in Jesus' time believed in reincarnation*]

"But what about you?" he asked. "Who do you say I am?"
Peter answered, "God's Messiah."
And he straitly charged them, and commanded *them* to tell no man that thing

[**Jesus was asking these questions in order to establish that he was the Messiah predicted in the Jewish Book of Prophets.**]

And in Matthew 17:10-13:

And the disciples asked him, saying, "Why then do the scribes say that Elijah must come first?" But he answered them and said, "Elijah indeed is to come and will restore all things. But I say to you that Elijah has come already, and they did not know him, but did to him whatever they wished. So also shall the Son of Man suffer at their hand." Then the disciples understood that he had spoken of John the Baptist.

[*A clear reference to the reincarnation of Elijah as John the Baptist. And the statement that "Elijah must come first" (before the Messiah) is referring to the prophecies in the Tanakh.*]

In the Old Testament (Extracted and translated from the Hebrew of the Jewish Tanakh):

And fifty men of the sons of the prophets went and stood to view afar off: and they two stood by the River Jordan. And Elijah took his mantle, and wrapped *it* together, and smote the waters, and they were divided hither and thither, so that they two went over on dry ground.

And it came to pass, when they were gone over, that Elijah said unto Elisha, "Ask what I shall do for thee, before I be taken away from thee." And Elisha said, "I pray thee, let a double portion of thy spirit be upon me."

And he said, "Thou hast asked a hard thing: *nevertheless*, if thou see me *when I am* taken from thee, it shall be so unto thee; but if not, it shall not be *so*." And it came to pass, as they still went on, and talked, that, behold, *there appeared* a chariot of fire, and horses of fire, and parted them both asunder; and Elijah went up by a whirlwind into heaven. And Elisha saw *it*. (Second Kings, 2:9)

[*Elijah was reincarnated as John the Baptist. Could it be that Elisha, with twice the spiritual power as Elijah, was reincarnated as Jesus?*]

In Malachi, the last book of the Old Testament, we find the prediction:

Behold, I am going to send you Elijah the prophet before the coming of the great and terrible day of the LORD. (Malachi 4-5)

[It is no coincidence that the Jordan River played a central role in both the elevation of Elisha by Elijah and the baptism of Jesus by John the Baptist. In the scriptures, the River Jordan symbolizes the crossing of a barrier and the transcendence of Spirit over matter.]

In addition to the texts that became the Christian Bible, texts written around and shortly after the time of Jesus by a group called "Gnostics" also recorded the sayings of Jesus, but because they contained some things the religious institutions of that time did not want propagated, they were banned as either too Jewish on the one hand, or too Christian on the other.

Gnosticism (from Ancient Greek: γνωστικός, *gnostikos*, "having knowledge", from γνῶσις *gnōsis*, knowledge) is the name given to the ancient religious ideas of Jewish-Christian groups in the first and second century AD. The earliest Christian sects, from the time of Jesus, believed in the Gnostic doctrine of emanation from one eternal Source: the idea that all individuals have their origin in God, and all have, in their inmost being, an eternal spark of God. They believed that the material world is created by emanations from God, and that there exists within each human body a Divine spark that can be gradually liberated in the course of many lifetimes by the attainment of gnosis, i.e., knowledge.

Gnostic Christians taught that periodic saviors of the world, from Krishna to Christ, were able to rekindle the divine spark in those in whom it had gone out. But organizers of the political doctrine that became the basis of the Catholic Church held that man was created by God as a physical being, *not a spiritual one*. Man, therefore, has no intrinsic connection to God, no divine spark through which he can reach God, according to Church doctrine.

Rejecting the part of original Christian thought that held that the soul is spiritual and immortal, an idea that some Church Fathers like Clement of Alexandria came back to, these early misguided Church Fathers developed the concept of *creatio ex nihilo*, or creation out of nothing. A human being could never be on a part of God, they reasoned, therefore his or her soul could not be part of God; so, it could not have been created out of God's essence. God must have created souls, along with the rest of the material universe, out of nothing at all. As Tertullian described it, God "fashioned this whole fabric with all its equipment of elements, bodies, spirits... out of nothing, to the glory of His majesty." The soul thus has no part of God inside itself.

Thanks to the Emperor Justinian, and the Catholic Church that accepted his decree designed to suppress ideas that elevated the potential of the consciousness of human beings to cosmic consciousness, that doctrine persists to this day. The *New Catholic Encyclopedia* says:

"Between Creator and creature there is the most profound distinction possible. God is not part of the world. He is not just the peak of reality. *Between God and the world there is an abyss....* To be created is to be not of itself, but from another. It is to be non-self-sufficient. This means that deep within itself [the soul] is in a condition of radical need, of total dependence.... It means to accept the fact that the world has no reality except what the Creator thinks and wills." (Emphasis added.)

In other words, according to institutionalized Church doctrine, there is not, as the Platonists believed, a great chain of being linking the creation to the Creator and enabling the creation to return to the Creator. There is no divine spark inside each heart. God created everything to run on its own, without any further involvement on his part. This dogma allowed the Church to promote itself as the only path to salvation of the soul, and thus control the masses.

But, the enforced doctrine of the Catholic Church contradicts several Biblical passages like Psalms 82:6:

"I have said, Ye *are* gods; and all of you *are* children of the most High."

John 10:32-34: When a mob threatened to stone him,

Jesus answered them, "Many good works have I shewed you from my Father; for which of those works do ye stone me?"

The Jews answered him, saying, "For a good work we stone thee not; but for blasphemy; and because that thou, being a man, makest thyself God."

Jesus replied, "Is it not written in your Law: 'I have said you are gods' ?

Jerimiah 1:4-5: Concerning the pre-existence of souls:

Then the word of the LORD came to me saying:

"Before I formed thee in the belly of thy mother, I knew thee; and before thou camest forth out of the womb I sanctified thee, *and* I ordained thee a prophet unto the nations".

Conclusions Regarding the Reincarnation Hypothesis

The weight of evidence, in terms of the number of highly intelligent people, past and present, who believe in reincarnation, and the numerous cases of remembered past lives with validated details, far outweighs the anti-reincarnation dogma of Islam and the Catholic and Protestant Churches that are based on non-Christian ideas institutionalized purely as social and political agendas.

The idea that there might be a certain finite number of souls stems from the illusion of finite forms with beginnings and ends. Because we become identified with finite physical bodies, and inhabit them for a while, physical bodies that are born, grow, decay, and die, we forget that we are immortal souls. While, because there are a finite number of organic life- supporting bodies at any one given point in time, and there are a finite number of souls incarnate on the Earth, we forget that the matrix of Primary Consciousness from which we come is infinite, and can project any number of souls into the physical universe at any given time. But, is all this proof that you or I, or any person currently alive has actually lived before? No, but it is an argument that it is possible.

An Overview of the Reincarnation of Souls Hypothesis

Reading the writings of reincarnated souls that we have not met in this lifetime, and talking with those that we have met, we can begin to piece together a picture of how reincarnation works. The mechanism and driving force behind it is spiritual evolution, not physical evolution, and the logic of the mechanism is as mathematical and irrefutable as the dimensionometric mathematical logic that reveals the existence of gimmel, the spiritual aspect of physical creation.

There are 'old' souls, as we saw above, like Paramahansa Yogananda, Mark Twain, Salvador Dali, Edgar Cayce, the Dahli Lama, and there are 'younger' souls, like Sylvester Stallone and George Harrison, and there are many, many 'young' souls who do not remember past lives at all. The terms 'Old' and 'young' in this context, do not necessarily relate to time spent on this Earth, but to lessons learned. It appears that progression and regression in a given lifetime - or lifetimes, are the results of the desires, choices, actions, and focus of the soul in question.

The law of cause and effect, or karma, as it is called in Eastern philosophy, largely governs the physical aspects of reincarnation, including the when, where and how of birth, social and economic position in life and death, time after time, but has no effect on the spirit, the essence of Primary Consciousness, which is the heart of the soul. Finite expressions of your unique experiences may carry over from previous lives into this one in the form of birthmarks and other physical characteristics. Evidence strongly suggesting this is found in the case studies of Dr. Ian Stevenson, and the documented readings of Edgar Cayce. Deeper soul characteristics, including the level of enlightenment attained, are also carried over, but re not necessarily displayed for all to see in this lifetime.

If, in a previous incarnation, you attained sharply focused awareness, involving a high level of intellect and a deep level of compassion and love, you will recognize friends and foes incarnate in this life. This can be helpful in maintaining and improving your focus and awareness amid the challenges and struggles of this life. People who have attained high levels of success in past lives are more likely to achieve success in this life also because, at a deep level, memories of purpose, focus, and methods for attaining alignment with the laws of the universe are still there.

Unfortunately, the same is true of bad habits. Just like physical patterns, psychological patterns persist unless steps are actively taken to change them.

Personal Experiences Suggestive of Past Lives

Disclaimer: *As a scientist, I am, by definition, a professional skeptic. This means that I will consider the transmigration or reincarnation of human consciousness, including my own, from a past living body or bodies into current living bodies a hypothesis, something that may or may not have happened. As far as I am concerned, it must remain a hypothesis, something to be proved or disproved, until there is indisputable evidence one way or the other.*

I will, on the other hand, stand behind the declaration of the existence of the third form, not measurable as matter (mass) or energy, because it has been proved to be true with mathematical logic and the empirical evidence of scientific data. This discovery supports the hypothesis of the survival of consciousness by extension of the law of conservation of mass and energy to include all three forms of the substance of reality. Certain experiences and memories may be strongly suggestive of reincarnation, but it is possible that they might be explained in other ways as well. For example: memories that appear to be from past lives could be somatic, i.e., memories of ancestral lives recorded in the DNA, that surface in the brains of descendants.

I have had a number of distinct personal memories and experiences that are suggestive of past lives, some of which I wrote about in "The Book of Atma" Published in 1977. My memories are suggestive of at least seven distinct lives in specific past time periods. Some have been validated by physical evidence, relevant information and unusual experiences in this lifetime. Still, they could be somatic memories of other human beings recorded in my DNA, that somehow, my conscious mind has tapped into. But I don't think so. If asked what I believe regarding my experiences suggestive of reincarnation, I will have to say that I believe that my soul did not appear from nothingness on October 7, 1936, and will not disappear anytime in the future, because there is no evidence that *anything* appears and disappears absolutely, it only changes.

All of the laws of nature discovered so far, especially the laws of conservation of mass and energy, tell us that nothing is ever created from nothingness or destroyed absolutely, and that all things change and evolve. The idea of the existence of nothingness is completely illogical and not supported by the evidence of scientific data. The discovery of gimmel strongly suggests that the same is true for consciousness. It may change and evolve, but there is no basis to believe that it will ever cease to exist. In addition, the existence of the complexly ordered physical universe only makes sense, and has purpose and meaning, if the progress made in one life actually carries over in some form into the next life. I am convinced by the evidence of experience and mathematical logic that there is something instead of nothing because there never was a state of absolute nothingness. Without gimmel, the laws of mechanics and the second law of thermodynamics tell us that no structured universe consisting of spinning objects like electrons, protons and neutrons could ever exist.

The fact that I exist now as a conscious being, implies that I have always existed, and will always exist in some form, as indicated clearly in the Judeo-Christian scriptures, especially before they were redacted by the heavy hand of Justinian I, Emperor of Rome.

PART VII: CONSCIOUS EXPERIENCES OUTSIDE THE BODY

Personal Experiences

While personal experience is considered to be subjective and of no scientific value by mainstream scientists, personal experience is actually the only direct connection we have with reality. When experience is pure, and unfiltered by dogmatic beliefs and borrowed philosophical or religious theories, it is real, and it is a window into the logical structure of Infinite Reality. The logic of mathematical theory is not always necessarily congruent with the logical structure of the cosmos, but when it is, everything comes into focus. and things that are inexplicable in the current physicalist paradigm of scientific materialism become clear. The following personal experiences are offered as evidence that consciousness is independent of physical bodies under certain circumstances.

Memories, Out-of-Body (OBEs) and Near-Death Experiences (NDEs)

The first OBE I remember in this life occurred before I was born: On October 7, 1936, I came into this world in a narrow green valley located between pre-Cambrian monoliths of porphyritic rhyolites, a few miles from the geologic center of a cluster of giant exfoliated granite boulders called Elephant Rocks, ringed by dark layers of basalt and other solidified flows of ancient magma. I remember the experience of rushing through a star-studded sky, as if drawn to this bulls-eye target of crystalline stone. This memory is as real as my memory of driving home from town yesterday.

While I may no longer have a photographic memory, I do have a number of clear, detailed memories of things that happened before I reached the age of nine months. One such memory is a vivid memory of an event that occurred when my parents moved from my grandfather Close's house, where I was born, across the street to the first house of their own. I described details of that move to my mother years later, when I was in my teens. She at first didn't believe that I could possibly remember anything from an event that occurred when I was less than nine months old, but, when I described numerous details of the furniture that was moved out before we moved into our new home, and features of the house that my father, a carpenter, changed during the first few months, she was convinced.

I can still see the scene as I saw it that day while being held in my mother's arms. I remember a dresser that had been painted dark blue. I can see the cracked paint and a rag doll hanging from the left upright holding the mirror, and I remember other details about other furniture and belongings of the previous occupants of the little three-room house. I watched as they were carried past me and out the back door of the house where a large square quarried stone served as a step into the back yard. My dad was helping the previous owner of the house load their furniture out onto a flat-bed truck with wooden side racks. Two or three years later, my father built another room onto the house, and that back door no longer existed. My description of the blue dresser and doll and the back door convinced my mother that I really did remember things that I witnessed before I could walk or speak.

I began speaking in short sentences at nine months; which I'm told, is somewhat unusual. But, perhaps more unusual was the fact that much of mathematics, especially algebra, was something I didn't have to learn, I just remembered it, prompting my maternal grandmother to say I was "too smart to live". I have memories of at least two lives as a mathematician, one in France and, more recently, one in Germany. I have memories of their names, their work and some of their experiences, and these memories are similar in quality to memories from my current lifetime.

Are pre-natal memories proof of the existence of my soul before birth and after the births and deaths of many lives? No, they are not proof, unless they can be verified, and verification is very difficult, if not impossible. Such verification of memories is difficult to obtain, even within a given lifetime, and it would most likely be even more difficult for memories of past lives with the trauma of physical death in between, for them to be verified. Consider, for example, the fact that I spent some time in Ft. Pierce Florida as a child.

My mother and I travelled by train from Arcadia, Missouri to Ft. Pierce, in 1945, to be with my father, who was stationed there as a Petty Officer in the US Navy. We stayed about a month before my dad was transferred to Treasure Island in the San Francisco Bay to prepare for the invasion of Japan. I remember the hotel we stayed in, the owner's Great Dane, who immediately became my friend when we met on the night of our arrival. I remember the "jungle" a few blocks north of the hotel, where I played. I remember the first palm trees and mango trees I had seen in this life, and the pier where I enjoyed fishing. I remember drinking fresh orange juice, eating fresh delicious fried fish sandwiches in the little restaurant on the pier, and drinking the limeade my mom made from the limes I picked from a tree behind the hotel.

I have not been back to that location since 1945, and most likely, the hotel, the jungle, and the lime tree, are long gone. If I were to go there today, it is very unlikely that there would be any physical evidence whatsoever with which to verify my memories of the place. My point is that the environs of the life of the German mathematician that might corroborate my memories of that life would probably have changed even more than those in Florida with the passage of more than 200 years; so, it would probably be a waste of time to go to Wasungen and Giessen, Germany expecting to recognize things that would verify memories of that life, and the life of the mathematician in France was more than 300 years ago. While I might feel a familiarity or have a *de ja vu* experience, in Germany, France, the Holy Land, Egypt, Petra, Saudi Arabia, India or Tibet, such experiences and even visual memories, do not constitute scientific proof.

Documented OBEs are more objective, and therefore, better evidence of the existence and survival of consciousness independent of the physical body than pre-natal memories. I have experienced several OBEs during my life that make it clear to me that my consciousness has existed, does exist and will exist independent of my physical body. Several of them happened while I was in elementary school, but, at the time, I thought of them as dreams. The first OBE that convinced me that OBEs were real happened when I was about ten years old. It involved my uncle George Tyndall and a toy airplane. I will describe it in some detail later in this book.

I have already described the series of OBEs I experienced in the Great Pyramid in Egypt above in Part V of this book, so, I will not describe them further, except to say that I might have considered them to be hallucinations if I hadn't experienced OBEs as a child, and other people observed what I saw. But I think it is of interest in this discussion to mention that my wife Jacqui experienced the vertigo and some other aspects of the OBEs with me, thousands of miles away in Missouri. The link between us appears to have spanned both time and space in multiple lives. We specifically remember events of lives together in the Middle East, Arabia and Europe. These shared memories make it harder to dismiss such memories as somatic memories of biological ancestors in our DNA, rather than personal incarnations, but not impossible.

OBEs and NDEs Experienced by Others

Many books have been written and published about OBEs and NDEs by those who claim to have experienced them and even by doctors, psychologists and hypnotherapists who have studied them. And serious scientist are investigating and documenting reports of OBEs and NDEs now, more than ever before. Of course, reading about the experiences of others is not as real and meaningful as personal experience, but some are pretty convincing, especially if they are related by someone you know very well, or someone with whom you have an emotional or spiritual connection, like a wife, soulmate or best friend.

Jacqui

Jacqui and I have been together for more than forty-two years in this lifetime, and she has been a great support and helpmate for me as I pursued independent research into the relationship of consciousness to the physical universe, when discussions of non-physical phenomena were taboo, and few in academia were willing to venture beyond the limits of scientific materialism. Jacqui has a sharp eye for detail and has been a wonderful editor and critic of everything I have written during all that time. The circumstances of her early life prevented her from obtaining a college or university degree. Her mother committed suicide when Jacqui was fifteen, and she had to work in the family business. But she is a very able researcher in her own right, and as we moved from state to state and even to foreign countries, as was required by my career as an environmental engineer, she enrolled in colleges and universities and took many courses that were of interest or utility, without concentrating on a specific degree program. As a result, she has a better understanding of a wide range of subjects than most people even those with advanced degrees.

Jacqui has always had a desire to help others. She was one of the first fully certified CARE instructors trained by D. Gary Young and Dr. David Stewart. She taught CARE classes for several years (CARE is an acronym for the Center for aromatherapy Research and Education), she earned certification and recognition as a registered aromatherapist with the National Association for Holistic Aromatherapy (NAHA) and she taught continuing education classes at Southeast Missouri State University. She is loved by many she has befriended and helped over the years as a nationally registered aromatherapist, and the many people who have read the books we have co-authored together and attended our talks at seminars and conventions in alternative health care in a number of cities in the US, and Australia. She also hosted numerous online seminars informing people about the uses of pure essential oils for many years.

Jacqui is a remarkable person in more ways than one, not the least of which is the number of NDEs and OBEs she has had in her life. She has had at least ten NDEs, and I have direct personal knowledge of six of them. She has been struck by lightning five times, two times since we've been married. I joke that, if you are out in an electrical storm with the choice of standing near Jacqui or a tree, you might want to choose the tree! She had numerous OBEs, especially when she was young. She has told me about experiencing identification with the wind moving over the South Florida coastline, during which she could see her own physical body lying on the beach below. She also recounted OBEs during which she was painfully aware at times, without any volition on her part, of violent emotional events happening to others, including suicides and murders that she experienced as if they were happening to her. After we met, she asked God to remove such unwanted psychic abilities, but she is still very intuitive.

PART VIII: THE NATURE OF REAITY, ITS PURPOSE AND GOAL

The Goal of Sentient Life

I believe, based on experience and reason, that the attainment of Cosmic Consciousness, the complete oneness with Primary Consciousness, the substrate or matrix within which everything has its being, is the real goal of life. With that as our goal, we have an excellent reason to work to improve ourselves in this life, anticipating that it may be transferred from one life to the next, and that we may transcend physical life at some point. This gives us a valid reason to strive to master the problems and enigmas of physical, mental and spiritual existence. Such mastery should give satisfaction and enjoyment even greater than that experienced with the mastering of a musical instrument, or any creative expression, including pure mathematical logic and the scientific method. The real goal is not a "theory of everything", it is becoming one with everything!

Consciousness, Spirituality and the Cube

Focused intent, diligent application with perseverance until success is achieved, always bring gratification and satisfaction. Mastering the cube is no exception. The cube, because of its geometrically orthogonal construction, conforming to the logic of rational reality, represents the interface of consciousness and the physical universe, i.e., the interface of what is known as mind, and what we experience as matter and energy. The fact that the cube can be transformed from a jumbled, apparently meaningless, randomly scrambled state to a meaningfully aligned state by the application of specific algorithms consisting of specific sequences of rotations, makes it a remarkedly appropriate model of the structure and logic of reality at the quantum, mid-range, and galactic scales, and at the physical, mental and spiritual levels of the human soul.

Learning how to transform the cube from any of the huge number of possible scrambled states to the beautifully coherent state obtained when all of the individual units are aligned so that the sides of the cube are monochromatic, mirrors the process of the orderly transformation of the condition of sentient life forms from physical, mental and spiritual disorder and chaos, to a blissful state of conscious alignment with, and awareness of the Oneness of All Things.

The lesson to be learned from the mastery of the cube, or indeed, from the mastery of any complete expression of the potential present in the structures of the physical universe, involving all of the existential forms of reality, mass, energy, space, time, and consciousness, is that your individualized consciousness, encased in the flesh and blood of the vehicle of a living physical body, is capable of becoming one with the infinitely logical consciousness behind all forms. Once you become aware that *that* is the ultimate purpose and goal of existence, you are acutely self-aware and consciously propelled to learn how to align yourself with the natural laws of the universe, and you are on your way to enlightenment.

A scrambled cube is symbolic of disorder and chaos, while the solved cube with monochromatic red, white, blue, green, yellow, and orange sides is a symbol of the elegant beauty and order of Primary Consciousness. Learning the laws governing the sequence of synchronistic patterns that leads to the solved cube, is analogous to learning the patterns of actions in our lives on this planet that lead to enlightenment and Cosmic Consciousness. Just as there is an optimum path from any scrambled state of a cube to the solved state, and there is an optimum path from any state of human confusion to Cosmic Consciousness. If you are in a state of physical, mental and spiritual disarray, it is your job to restore order to your life, and find your optimum path back to a state of Primary Consciousness. Better known as the Kingdom of God. Like the singing of a hymn, the repetitive practice of the solution of the cube can become a prayer and a meditation.

Prayer, Meditation and the Conscious Universe

The discovery of gimmel reveals the fact that every atom of the physical universe is part of, and is connected with the infinity of Primary Consciousness. Prayer is the act of trying to communicate with Primary Consciousness. But the finite mind of an individual conscious being which cannot fully comprehend Infinity, relates to reality in terms of subject and object, self and other. So, traditional forms of prayer tend to consist of visualizing an objective form, such as an imagined Supreme Transcendental Being, or the image of an enlightened teacher, to whom our prayers can be addressed.

An effective prayer begins with an expression of acknowledgement of, and gratitude for the miracle of your existence, and a willingness to accept the will and purpose of Primary Consciousness. After expressing the purpose of the prayer in the form of a verbal description and/or visualization of the desired outcome, one should pause and expect a response. In this silent and receptive attitude, the supplicant should refrain from verbal expression or thought and try to feel the presence of Primary Consciousness. Primary Consciousness is omnipresent, sustaining the meaningful structure and purpose of physical reality and the coherent cosmos through the agency of gimmel, the connection with the Akashic field (see pages 78 and 79), which is filled with love and Compassion.

Meditative Communion with Primary Consciousness

The silent attitude of expectation is just the first step toward a true meditative state. Real meditation, i.e., communion with Primary Consciousness, begins when a connection with the Akashic field is established, and one feels the presence of Primary Consciousness as the "Holy Ghost", or Spirit of God. In the early stages of meditation, when the connection is being established, a token of this connection may manifest in your consciousness in the form of light and/or sound.

The light may appear as waves of white and blue light, sometimes referred to as the Dove of Peace descending into your consciousness. The sound may be heard internally as a deep pulsating vibration, like the rushing of flowing waters, or the hum of a flow of a powerful current. The light and sound together signify the presence of the Holy Spirit, referred to as the Holy Ghost in Jewish and Christian scriptures. These signs of the beginning of meditative communion with Primary Consciousness are not easily attained until an advanced state of spiritual awareness is reached. These encouraging signs are variously alluded to both directly and obliquely, in the scriptures of the major religions of the world. For example:

Judaism Tanakh, Samuel 6 & 7: 6.

And the Holy Ghost will descend upon you, and you
תִיָבַנְתִהְו זֹהִי חַוְר רְיֵל עַ הָתַ לְצָו ו

will prophesy with them, and you will another man.
be turned into
רחָ אַ שִׁיאְ לְ כַפְהַנ וְ מֵע:

7. And it will be, that when these signs will come to you,
הַנָיאֹבַ תָ ביתכ) הָנָאֹבָת יִכָ יָהָו ז)
do for yourself what your hand will find, for God is with you.
רְשַׁ אָ הְל הַשָׁ עַ לֵ הַל אַ הָ תוֹתָ אֹה
יִכָ דָ יְיָ אצַ מָ תִאצְמת

Christianity

"The light of the body is the eye: if therefore thine eye be single, thy whole body shall be full of light -Matthew 6:22

"In the beginning was the Word, and the Word was with God, and the Word was God." -John 1:1

Buddhism

I am free from space and time. Mine is the joy of the unclad...
My form consists of total light;
The light of pure consciousness am I. - the Maitreya Upanishad

The word which all the Vedas proclaim,
That which is expressed in every Tapas (penance, austerity, meditation),
That for which they live the life of a Brahmacharin Understand that word in its essence: **Aum! that is the word.**
*Yes, this syllable (**AUM**) is Brahman. This syllable is the highest. He who knows that syllable,*
Whatever he desires, is his. — Katho Upanishad, 1.2.15-1.2.16

Hinduism

God manifests in creation as the Cosmic Vibration, which expresses itself as Cosmic Sound and Cosmic Light. The Cosmic Sound or Aum is the synthesis of all the sounds of the highly vibrating life forces (lifetrons), electrons, protons, and atoms. By listening to Aum, the yogi becomes a true brahmachari or one who is attuned to Brahman. By deep concentration the devotee can hear Aum at any time and in any place. – Paramahansa Yogananda, Self-Realization Fellowship, *Yogoda Satsanga* in India.

The Range of Awareness: Minimal to Cosmic

The possible range of conscious awareness in human beings ranges from barely self-aware, to full cosmic consciousness. Relatively few individuals exist on the extreme upper and lower tails of the bell-curve (Gaussian distribution) of consciousness, and for the average person, at this point in the history of the human race, awareness depends primarily on the functioning of the physical organs of sight, hearing, smell, taste, and touch, organs that actually act as reduction valves that serve to prevent informational overload of the neural network systems of the brain with incoming data. As pointed out earlier, only relatively small portions of the ranges of vibratory energy that flood the universe are allowed into the imaging processes of the brain of the average sentient being. At the lower limits, the levels of awareness and cognition are just enough to make physical survival possible, and at the upper limits, there are only a very few individuals with total cosmic consciousness; they are fully-enlightened spiritual masters.

Spiritual masters are often aware of reality to a depth and degree difficult to explain in any human language, so they must resort to instructional analogies to convey their experiences in any meaningful way to the rest of us. The multi-dimensional vibratory energy centers known as *chakras*, for example, are likened to flowers with petals. They are not flowers, of course, but the flower image provides an analogy that anyone can understand. The average human being can understand, or at least imagine states of consciousness that exist slightly above or slightly below their own position on the bell-curve of consciousness, but states five or six standard deviations above or below, especially above, are almost impossible to imagine. Multi-dimensional global awareness, even over a limited domain of space-time, is hard to describe to anyone who has not experienced it. Here is one of the best descriptions I have found of an actual experience of cosmic consciousness. It was bestowed upon Mukunda Lal Ghosh (Yogananda's birth name) by his guru, Swami Sri Yukteswar. It is found in Chapter 14 of *Autobiography of a Yogi*.

> *He (Sri Yukteswar) struck gently on my chest above the heart. My body became immovably rooted, breath was drawn out of my lungs as if by some huge magnet. Soul and mind instantly lost their physical bondage and streamed out like a fluid piercing light from my every pore. The flesh was as though dead; yet in my intense*

PART VIII: THE NATURE OF REAITY, ITS PURPOSE AND GOAL 245

awareness I knew that never before had I been fully alive. My sense of identity was no longer narrowly confined to a body, but embraced the circumambient atoms. People on distant streets seemed to be moving gently over my own remote periphery. The roots of plants and trees appeared through a dim transparency of the soil; I discerned the inward flow of their sap.

The whole vicinity lay bare before me. My ordinary frontal vision was now changed to a vast spherical sight, simultaneously all-perceptive. Through the back of my head I saw men strolling down Rai Ghat Lane, and noticed also a white cow that was leisurely approaching. When she reached the open ashram gate, I observed her as though with my two physical eyes. After she passed behind the brick wall of the courtyard, I saw her clearly still."

All objects within my panoramic gaze trembled and vibrated like quick motion pictures. My body, Master's, the pillared courtyard. The furniture and floor, the trees and sunshine, occasionally became violently agitated. Until all melted into a luminescent sea; even as sugar crystals, thrown into a glass of water, dissolve after being shaken. The unifying light alternated with materializations of form, the metamorphoses revealing the law of cause and effect in creation.

The experience continued to expand and intensify:

An oceanic joy broke upon the calm endless shores of my soul, ... A swelling glory within me began to envelop towns, continents, the earth, solar and stellar systems,

> tenuous nebulae, and floating universes. The entire cosmos, gently luminous, like a city seen from afar at night, glimmered within the infinitude of my being. ...

In this fascinating account, the key to understanding the nature of Cosmic Consciousness relative to ordinary 3S-1t perception, is the phrase *"simultaneously all-perceptive"*. Think of the total awareness of Cosmic Consciousness as simultaneously all-perceptive, combining the global awareness of all the senses expanded to include all of the patterns of vibratory information flooding the universe, and think of ordinary awareness as a severely fragmented and limited outline of the omniscient awareness of cosmic consciousness. Some of the cognitive states experienced in OBEs, NDEs and other so-called "altered states of consciousness" provide brief glimpses of certain aspects of Cosmic Consciousness.

The Future of Science and the Training of Scientists

If mainstream science is to outgrow the limits of materialism, and become a truly comprehensive effort to understand reality, the training of scientists must go beyond the superficial training of the mental faculties that passes for education today. Instead of just a few years learning the basic methods of investigation used in the study of the small portion of reality known as the physical universe, budding scientists should also learn how to expand their consciousness to the point of being able to explore psychic and spiritual phenomena and their relationship to physicality. Without that, the chances of misuse of knowledge of some of the inner workings of the universe that can be detriment of us all, grows greater and greater with each passing year.

Scientists in all disciplines also need to have a deeper understanding of mathematical logic, dimensionometry (geometry expanded to include hyper-dimensional domains) and infinities. TDVP, with the calculus of dimensional distinctions, represents a major advancement in that regard because the CoDD incorporates all three. The history of the study of mathematical infinity illustrates the reluctance of mainstream scientists and mathematicians to accept infinity as a part of reality, something that has to be done, if consciousness is to be properly included in scientific investigation. Georg Cantor, a brilliant German mathematician who was a pioneer in set theory, also developed the concept of the infinity of infinities. Much like mainstream scientists reject consciousness as a fundamental aspect of reality now, the mathematicians of his time rejected infinity as a fundamental concept of mathematics.

PART IX: A COSMIC LOVE STORY

Two persons who unite their lives to help each other toward divine realization are founding their marriage on the right basis ... When man and woman genuinely love one another, there is complete harmony between them in body, mind and soul. When their love is expressed in its highest form, it results in perfect unity. – Paramahansa Yogananda

The Undying Love of Quantum Entangled Souls

Like the tri-axel heart of the cube, Divine Love holds the universe together. Like duct tape, the love that brings two lovers together in this world, has a light side and a dark side, but spiritual progress is obtained for both lovers when the focus is on the light side, the side that reflects God's love for all souls. Does true love transcend the grave and grow as an eternal flame from one life to the next? Yes, I believe it does, and I believe that the memories I share with Jacqui are real evidence supporting that belief. When we met in this life, it was like a continuation of something wonderful.

All souls are entangled, but we are not fully aware of that truth until we are far advanced on the path to spiritual enlightenment. Ultimately, all souls are one in Divine Love. This is reflected in the infinitely expanding cosmos and in the finite world of everyday experience, in the nature of consciousness. To function in this world as physical beings, we must first divide reality up into two parts: the observer and the observed, and then into three and more parts, as we draw distinctions in the observed part. Consciousness is never experienced in the plural, even when we are identified with a finite physical body. There is only one self-referential reality that is eternally consistent. This is what makes the scientific study of the experience of consciousness, love and spiritual enlightenment so difficult. Advanced mathematical physics is child's play by comparison.

Physical reality is finite and thus ostensibly limited. But, even within the illusion of linear, unidirectional time, it is potentially infinite because it is expanding into infinity at the speed of light. The finite cannot contain the infinite, but the purpose of this part of the book is to open the door for a glimpse into the infinite potential of spiritual experience by sharing our memories of this and some of the previous lives we've shared.

Struck by Lightning, - and Cosmic Energy!

> "I am what I shall be, and only time remains the distance."
> ~~ Jacquelyn Ann Hill, 1971

It was the summer of 1958, and in a small town in Middle Tennessee, a five-year-old red-haired little girl was playing happily in her parents' driveway on her brand-new bicycle. Her cousin and her younger brother were riding in circles with her around the end of the driveway nearest the garage. Suddenly her hair stood on end, on both her head and her arms like the seed-bearing fibers of a dandelion, crackling with static electricity. There was no sky-splitting crack of lightning. It was silent, but she felt the lightning coursing through her body, surging from the ground beneath her feet, into the sky above. The next thing she knew she was laying on the ground and seeing angelic beings of light and hearing an angelic choir singing hymns to the Holy Father. The Light became a single point and then expanded into a great ocean of light that called out to her in loving tones. Oh, the love that coursed through her, she had never known such love. This love was so different than human love, so full and open and unconditional. She wanted to stay with this LOVE forever.

Just as suddenly as the lightning had taken her to this place of Love, her mother's screams "Jacqui! Jacqui! Please wake up!" brought her back to earth and awareness of her surroundings. The angelic host was gone. Oh, but little Jacqui did not want to wake up from this whatever it was, a dream? It felt so real, more real than any experience she had ever had.

"Mama, it's so beautiful. I want you to see what I saw and hear the beautiful music," she said out loud. But that did not happen.

Her mother gathered her up in her arms and carried her into the house, crying tears of joy that her little girl had been spared. She was alive. Jacqui was confused. The experience had suddenly ended just as quickly as it had started, and it left a hole in her heart that only that LOVE could fill.

A tiny little girl, only 5 years old, had seen and heard God speaking in Light and LOVE. How could she get back there? This was all she wanted to do, but it did not seem to be an ability she had. Only Jesus could bring her back there, she determined, but could she do something, anything to help her get back there? This was the greatest experience of her young life. Her brother and cousin did not know what had happened. They saw Jacqui's hair standup on end. They saw her fall over in the driveway, but they were unharmed and unaffected by the lightning. They had seen what happened to little Jacqui, but they didn't understand. No one understood.

Jacqui survived the lightning strike, and it was only one of many near-death events she was to experience during her lifetime, and this is her story. From that day forward, she was tremendously different. So different, no one could understand.

Was this a random event? Was it purely an *accident* that at that exact time, Jacqui was at the exact location when and where an electrical discharge between earth and sky just happened to occur? So, it seems. But what if that is not the case? What if what appears to be a random event is, in fact the result of a cause and effect pattern that is mathematically related to other events that occurred in the past and will occur in the future? Could it be that everything we experience has meaning and purpose, and there really are no accidents?

Jacqui was a very bright little girl, but after this experience, her consciousness seems to have expanded because she began to know things no little girl was supposed to know. Sometimes she knew what others were thinking before they spoke, and she knew about events that happened miles away. At first, she thought everyone could see and feel things the way she did, but when she told her parents about things she saw and knew, they did not understand; they were frightened and worried that their only daughter might be insane. For most Christians in the Bible Belt of the US, in the 1950s, people having the kind of experiences she was describing were thought to be evidence of possession by evil spirits. She learned not to talk about her out-of-body experiences (OBEs), and that made her feel isolated and alone. It would be many years before she would meet people who would understand. Were her OBEs caused by the lightning strike, or were they and the lightning strike just part of a pattern of awakening to a more comprehensive experience of consciousness?

For most human beings alive today, it seems that we are born into this world in the middle of a complicated story of which we know nothing. We arrive in a state of bewilderment, not knowing from whence we came, why we are here, or where we are going. And when we meet someone who claims to know more about this existence than we do, we think they are strange. But, what if the level of awareness considered "normal" is an illusion? What if our lives are actually part of a greater plan, part of an endless chain of cause and effect, part of the complex pattern of reality, largely hidden from us? Is there any evidence that this might be true?

Yes, there is: We know now that the physical senses, thought of as conduits through which we experience reality, are actually reduction valves. They allow only a very small portion of the vibratory energies radiating throughout reality into the central neurological processing system, the extensive network of axons and synapses in the brain. Multidimensional feedback loops allow the brain to compare the quanta of energy routed from the senses with logically recognizable images that exist in consciousness. What if the logical three-dimensional mental images in individualized consciousness are all preexisting in some form in what could be called Primary Consciousness, and in this sense, there is nothing new in the universe, and the illusion of the passage of time is created by the changes that occur in the brain.

Careful investigations reveal that the universe is related to consciousness like a time machine; a time machine even more precise than a Swiss watch, with most of its workings hidden. We only see a very small slice of reality, but scientific investigations of physical phenomena, from the rising and setting of the sun to the spinning of the electron, and everything in between, continue to uncover deep connections in the form of quantum entanglements. What if the same is true of consciousness and spirit? What if our lives as sentient beings are mostly hidden from us, like the workings of the universe, because of the limitations of the human brain and the related physical senses? Perhaps every soul is connected in the space-time-consciousness of the Greater Reality, and *only time remains the distance between your soul and every other soul,* --and the Infinite.

Six Hundred Miles Away: A Timeless Link

> Consciousness is three, as is Time and Space
> In nine-D Reality, we can clearly see:
> Everything exists at once, in the same place!
> - Edward Roy Close, December 2018

During the summer of 1946, a nine-year-old boy was fishing from the bank of a spring-fed creek on the edge of a small town in Southeast Missouri. He had walked across a field to his favorite fishing and swimming hole, about one-quarter mile from his parents' home. After catching several pan-sized fish, he decided to get some relief from the increasing heat of the afternoon sun by taking a dip in the inviting cool, clear water. He folded his clothes and placed them in the hollow of a large sycamore tree on the creek bank, because it looked like an afternoon shower was brewing in the west. He plunged into the water and swam the length of the pool a few times, and then stopped and stood still in the cool water. When his feet touched the bottom, only his head was above water. The cool water enveloping him from the soles of his feet to his chin was refreshing and comforting.

What happened next is forever etched in his memory: "I stood spellbound for an instant, as a throbbing surge of electrical energy hummed through the water and my body. Instantly, my awareness was expanded to include the whole body of water. I could feel the fish in the water, the banks of the pool and the roots of plants reaching into the water-saturated soil. This expanded vision was virtually simultaneous with a searing bolt of lightning that struck the top of a tall sycamore on the creek bank. With great destructive power, it ripped down the side of the tree, through the roots, into the creek. Fingers of bark and soil were thrown into the air, and a deafening clap of thunder followed the flash of lightning almost immediately."

Eddie remained standing in the water for several minutes, marveling at what had happened. The storm passed quickly, and he retrieved his dry clothes from the hollow tree and the stringer of fish from the creek. Walking across the field, he chuckled to himself. His mom would wonder how he had remained dry in the downpour that had swept across the valley. He decided, however, not to tell her about the lightning, and made a mental note that swimming in a creek lined with trees during a thunderstorm was probably not a good idea.

After that experience, I seemed to be more fully aware than before. Before that, riding my bicycle, reading and trading comic books had been my favorite pastimes. After my electrifying experience, I began reading books and magazine articles on a wide variety of subjects. I wrote a book of poems, started learning German and Spanish, and took an interest in things like electronics, fruit tree grafting, hypnosis, exercise, and health in general. My expanded consciousness changed the nature of my experiences during my waking hours, and even the nature of my dreams at night.

One weekend I was staying overnight in my maternal grandparents' home, a two-story farmhouse built by my grandfather around 1900. Sometime after going to sleep in an upstairs bedroom, I woke, finding myself standing in the middle of the large room that served as both living room and dining room on the first floor. I thought I must have been walking in my sleep. That was something that happened to me occasionally during my pre-teen years. When I started to move, to my surprise, I began to float gently upward like a balloon, and as my head was nearly touching the ceiling, I noticed a small yellow and blue toy airplane lying on the top of my grandmother's tall china cabinet near the door to the kitchen and across from the door to the stairs leading to the upstairs bedrooms. To my even greater surprise, my upward motion didn't stop when my head touched the ceiling! I seemed to continue to float upward, through the ceiling, through the upstairs bedroom next to the one where I slept, through the roof and out into the starry night. Flying across the valley toward a rising full moon, I could see the soft glow of kerosene lamps in some of the farmhouses below.

The next morning, my uncle George and I got up early to go fishing in a large farm pond about a mile away, but as we came downstairs, I remembered the "dream" and wondered ... could there actually be a toy airplane on top of the china cabinet? I dragged a chair from the table over to the china cabinet, and was starting to climb up onto the narrow shelf above the silverware drawers, in front of the glass doors of the china cabinet, when my uncle said: "What are you doing?"

"I think there's a toy airplane on top of the china cabinet."

"Get down from there, before you fall." He demanded. Being older and taller than me, he could reach the top by standing on the chair. Feeling around, he retrieved something and handed it to me. It was the blue and yellow toy airplane I had "seen" in my "dream"!

"Don't throw it up there again." He admonished.

But I hadn't thrown it up there. I'd never seen it before. It could have been one of my cousins, I thought, but I didn't say anything,

My senses seemed to be heightened after the shocking experience in Kuhn's Creek that summer afternoon. One morning, sitting in a fifth-grade classroom, looking at the teacher standing at the front of the room, my vision was suddenly transformed: The teacher's face seemed to fill all space. I could see the pores of her skin as if I were looking at them under a microscope, and I could hear her thoughts! The next minute, she seemed to zoom away into the distance, her head barely the size of a pea in my expanded field of vision.

At night, lying in bed before going to sleep, my auditory sense was transformed: I could hear the slightest noise. Any movement in the sheets sounded like an avalanche in the mountains. I could hear my father's pocket watch, lying on a dresser in my parents' bedroom, three rooms away, each tick resounding like a pounding sledge hammer. I found that I could drown these local sounds out by tuning in to music that seemed to be floating in the air. I would often drift off to sleep listening to a marching band, an orchestra, or a string quartet. I could hear every note with clarity, identify the instruments, and hear counterpoints and harmonies in great detail.

Could there be a connection between these two consciousness-expanding lightning-strike events involving Jacqui and me, even though they were miles and years apart? It wasn't until almost 20 years after Jacqui's experience in 1958, that we met in this life. We met as adults in Tampa Florida, a city meteorologists call the "lightning capital of North America." In an article posted on www.weatherbug.com July 29, 2018, listing the top 30 cities in the US with the most lighting strikes, Tampa stands out as number one, with the greatest number of lightning strikes per year.

Were these events accidents? Coincidences? Or were these experiences part of a network of mathematically related events in the real fabric of consciousness underlying physical reality?

The Meeting in Tampa

When Jacqui and I met in Tampa, something electric definitely happened. Here is how Jacqui describes our first meeting, in her own words, with some background information about events in her life that led up to it:

> After my vision of Christ as a child, and the experiencing of the indescribable joy of being enveloped in his all-encompassing love, I knew that my life would be about seeking the complete, unending ecstasy of that love again through serving others. My grandmother Hill was a pianist and church organist, and I knew that music could lift me up spiritually. So, after the failure of my first marriage, I decided to move to Nashville, only 30 miles west of the town of my birth, Lebanon Tennessee, and work in the music industry as a singer and songwriter.

Pretty young girls are always welcome in the music industry, and I knew that men considered me attractive. But I also worked hard at training my ear and voice by singing into a tape recorder in the privacy of my apartment, playing it back over and over, making sure it was musically sound, and that it conveyed the sincerity of my heart and soul. By frequenting Nashville's Music Row bars, I soon made contacts with people who could put me in touch with the movers and shakers of the country music industry. Within a few months, I was welcomed like an insider in some of the top recording studios in Nashville.

I traveled up and down the East Coast for a time, as the lead female singer in a band, to get some experience singing in lounges and night clubs and I returned to Nashville in 1973, and was offered a contract with a major recording company. I was on the threshold of a career every young singer dreams of: a pathway to fame and fortune. But I had had a glimpse of the dark underside of the music industry: many young singers, especially girls, embarked on lives of alcohol, sex and drugs behind the scenes in the industry, lured by the fantasy of fame and fortune. Some were told they had great talent and were signed to a contract, only to remain "on hold", as the owners and producers of the company that signed them promoted someone else as the recording star, for the entire period of their contract. This was a trap I wanted to avoid.

As I considered signing the ten-year contract, I had another prophetic dream: in it, I was told that, if I remained in the music industry, I would not live past the age of 24. I also knew that my father was an alcoholic, and that I had inherited the DNA of an addictive personality. I decided that the God who had blessed me with a glimpse of His glory, would not want me to go down that path. I was directed in dreams to leave the music industry and move to Florida.

Many times, both before and after I moved to Florida, I had dreams in which I saw a man with white hair and beard. I had seen this man in a dream over and over again ever since I was two years old. I was always told by a voice I knew to be God that this was the person who would **take me to where I was supposed to be, and that I would help more people than I could ever imagine!** I was also shown a young boy at about age 5. And I was told that I would be married to this man, and we would have this little boy, whose soulful dark eyes and dark complexion didn't look anything like me or this man. He looked like he belonged to someone else, but he also looked a great deal like what my father might have looked like as a child. So, I never questioned that this would be true. It was inevitable. God never lied. But I had never met this man, and I had already been married once and divorced for nearly 3 years when I moved to Tampa, following the messages I received in these dreams.

A friend of mine from the music industry offered me a temporary place to stay in Tampa, with no strings attached, for a few days. I met him at the address, and when we walked into the apartment, the phone began to ring. He told me that he had just had the telephone installed, and had not given the number out to anyone.

He picked up the phone and answered it. He turned toward me with a very puzzled look, and said: "It's for you!!!"

I said "Hello? this is Jacqui." A woman's voice on the other end said:

*"I'm a member of a spiritual group in Texas. I was told to call this number and tell you that you must trust your spirit guides. **If you do not despair, you will live to become a blessing to millions of people. You will help more people than you ever dreamed possible.**"*

I was stunned! Who was this on the other end of the line? How did she know how and when to call this number. Before I could find the words to ask these questions, the line went dead. I thanked my friend for offering this place to stay until I could find an apartment.

A few days later, another friend told me about the town of Casadaga, known as a "spiritualist camp", about 120 miles from Tampa, northeast of Orlando. This friend had had an amazingly accurate reading from a psychic who lived there. That interested me because of my recurring dreams and this strange phone call. I believed the dreams of the man who would take me where I was supposed to be, where I would help many people, were true and prophetic. Perhaps a professional psychic could help me find that man!

Casadaga was a quaint little town full of mostly small cottages, many with signs that had "Psychic Readings", "Palmist", or "Psychic medium" under the psychic's name, that almost always had the title "Reverend" in front of it. Rev. Mills was a middle-aged man with a serious, professional manner. Without any information from me, he confirmed that I would soon meet a man, "a teacher" of some sort, he said, who would "take me where I was supposed to be." But the only advice he offered was to go back to Tampa and wait. I was disappointed.

Life was O.K. in Tampa, but I was still living pretty much as I had lived in Nashville. The habits I acquired in Nashville were hard to break. The trip to Casadaga and the strange phone call validated the dreams that had haunted me since childhood; but how was I supposed to find this person who would take me where I was supposed to be? Through more dreams, God led to look for a job at an engineering firm located near the airport in Tampa. I had applied to an architect's advertised position for a secretary, but after 3 interviews, the woman who was my contact at the architectural firm said, "I don't think you're quite right for this job, however, I know someone who is looking for someone just like you. It's a small company near the airport. Would you be open to going to interview with them?" "YES! Yes! I would be happy to do that."

My heart soared with recognition that this was the job that would lead me to the man in my dreams. At the first interview, I was ready to go to work, and I made the mistake of saying, "I know this is my job. Can I start today?" I knew this was 'my job', but, unfortunately, they were scared by my enthusiasm for the job, and didn't hire me on the spot. They said they had others to interview and they would get back to me within the next week. So, I went home and waited for them to call. I waited, one week, another, and then another. No word, but I didn't go another interview. I knew this was the job. God had confirmed in my dreams that this was indeed my job.

I enjoyed the time by spending my days at the pool in the apartment complex where I had taken a small one-bedroom apartment. My friends here in Tamps thought I was crazy for not continuing to look for a job. They strongly urged me to apply for other jobs, but I wouldn't budge.

I was almost at the end of the month with no money to pay my rent on the 1st. God would not fail me, but WOW! Why was it taking so long to get started with this job? It was so confusing, but I trusted God and continued to wait. Finally the call came.

"Uh, are you still available?" Jerry Seaburn asked me.

"Yes, of course! When would you like me to start?" I responded.

"Can you start tomorrow morning?" Jerry sounded tired.

"Absolutely!" I said. "I'll be there at 8, unless you need me to come in at a different time."

"Eight is good," Jerry said.

"Great! Can I ask you one question?"

"Sure."

"Why didn't you call sooner?"

"We hired someone else. She just stopped coming in and hasn't called or anything." Jerry was exasperated, and it came through loud and clear.

"I see," I said. "Well, I'll be happy to come in today if you like."

Jerry responded, "No, that's not necessary. We'll look forward to seeing you tomorrow morning, bright and early. See you then."

The job was easy for me and yet I still found it exhilarating, because I knew this was the right job and it would lead me to the man of my dreams who would take me where I was supposed to be. But for some unknown reason I had not yet met the man of my dreams. Several months passed by, but I had been led by dreams from God to this job, and I knew somehow this job would lead me to what I was looking for, and it wasn't being married, it was something to help me get closer to God.

After several weeks, I began talking to a couple of the three people (the 2 owners, Jerry and Al, and one Geologist) who worked there about my beliefs. It was just a way to get more acquainted and be open and honest, but my open honesty was not always appreciated. I was talking with Jerry, one day, He was the principal

I felt most comfortable with, but he didn't seem very comfortable with me. Telling Jerry some of the things I believed in led him to say: "You have some pretty weird ideas. You need to meet a friend of mine, Ed Close". When Jerry said that name, in my mind's eye I saw the gray-haired, gray-bearded man I had dreamed about for so many years. It was as if there were a little picture in light right in my mind's eyes. I said "Oh Yes! I definitely want to meet that man! Can you set up a time for me to talk with him?"

Jerry looked at me sideways, as though he was shocked and a little scared by my question. "Ah, maybe. I'll see. Let me get back to you about that."

Another month passed. No meeting had been set up. One Friday, I became quite depressed that nothing was happening. All the pieces were in place. Why was God making me wait? What did I need to do?

I had given up the career I had wanted in the music industry, - for what? For dreams of a life that would be full of the Love of God and something I couldn't understand was why this was being withheld from me still. The same problems and temptations I had in Nashville, had followed me to Tampa! The 'job' I had been led to, had first gone to someone else. Maybe I was wrong. Maybe I was always wrong.

Then, I suddenly remembered having had vivid visions of Christ as a child, and wanting to be baptized, but my parents said I was too young. Remembering those visions, and the love I felt radiating from Christ, I decided to try to break the pattern of the past, that included both addiction and suicide, which seemed to run in my family, and already had roots in my life. Instead, I would spend the next few days crying and praying to Jesus. I called in sick and didn't go to work. Instead, I stayed locked in my little bedroom crying and praying.

What had I done wrong? What did I need to do? Why was this being withheld from me? God, you must come and answer my questions. I can't keep going like this. Life isn't worth it if I cannot be closer to you and that perfect LOVE you let me feel as a little girl. You have to come. You have to respond.

I skipped breakfast; I wasn't hungry. I continued crying and praying, vowing that I would not stop until I received an answer, - or die. One day passed. Two days passed. I only drank water and went to the bathroom when necessary. Every other minute was spent crying and praying to my beloved Lord Jesus Christ. He was all I knew and all I could know with certainty would not lead me astray.

Finally, during the late afternoon on the third day of fasting, crying and praying, I received an answer.

I heard God's voice emanating from the ceiling and everywhere all at once. "Do you trust me?" God asked.

Oh, my beloved, of course I trust you. I can do nothing else. "YES!"

Instantly I knew that everything had changed, and suddenly I was enveloped in a peace that surpasses all understanding. I felt that incredible LOVE again for the first time in so many years, it had begun to seem like a distant memory. Oh, but instantly I was carried up into heaven and once again felt communion with my beloved Lord. It was overwhelming and filling and so richly aware of God's LOVE that nothing else mattered. And I knew that I would be either meet this man within two weeks or I would be dead and in heaven. I didn't care which happened. I just wanted to be with God and always live in this unmatched LOVE that poured over me right at this moment. Then I feel into a deep sleep.

Monday morning came bright and filled with sunshine. I called and let Jerry know I was coming in to work. He asked if I could I go by the airport and pick up the mail at the Company's mailbox and then stop by the USGS to pick up some maps on my way in to work. Jerry said a lady named Dorothy would have the maps waiting for me at the front desk at the USGS office.

When I walked in the door, there was no one at the receptionist's desk. I looked around and suddenly realized that I knew every desk and chair and every picture on the walls. It was as if I had seen this place a million times before. I knew that this *de ja vu* experience meant only one thing: something special was about to happen! Then he walked around the corner, and said: 'May I help you?' My eyes nearly popped out. I said: 'You're Ed Close!' He looked at me closely, and said: 'Yes, I am. Do I know you?'

Within a few minutes, while he walked me around the office introducing me to various people, we were talking about meditation and lamaseries in Tibet. He confided to me that he regularly attended group meditation in Tampa on Thursday nights. I asked if he would take me to one of the meetings, and he said 'I don't usually take someone I've just met. Can you sit still and make no noise for an hour?' 'If that's what I need to do, I'll do it!' I replied.

That Thursday, when I went to my first meditation meeting, I met some other members of the meditation group, including Lana, who was to become my best friend. I knew that these people were old friends. I had known them in past lives. After a brief explanation of some simple meditation techniques, we sat down and listened to a devotional chant, I closed my eyes and floated away. When I opened my eyes, I realized that we were sitting in front of an altar, with Christ as the central figure, flanked by teachers of Self Realization Fellowship. I closed my eyes again, with tears of joy running down my cheeks. All space seemed to be filled with light and the sound of the *aum vibration*. I knew I had come *Home*!

My Memory of Meeting Jacqui

In Tampa, I was unhappy in my job and unhappy in my marriage. Pressures mounted, and my life became intolerable. After much soul-searching, I decided to terminate both my job and my marriage. Was that the right thing to do? Was it simply mid-life crisis? Was I going insane? It didn't matter. I was slowly dying inside, and I had to do something about it, so I left my job of 12 years, and my wife of nearly 18 years, and two adopted children. Was it the right action? Or was it terribly wrong?

Was I just running away, causing others pain? All I knew was that I had to do it, and once I had made the decision, there was no turning back.

Looking back, I realize that this crisis was mostly of my own making. If I had not turned back from the spiritual journey that I had set out upon in 1958, when David Stewart and I left College and found Self-Realization Fellowship, none of this would have happened. Like a wrong rotation of the cube, one decision can lead to multiple situations requiring many hard decisions, until the right one is made.

How did it come to this? I had made the decisions that created the situation. I was trying to live two different lives at the same time. I had entered on the spiritual path again, in 1960, just after we were married. I was fully committed to the path, but my wife was not, and that was the root of the problem. We were drifting apart, Bobbi had a traumatic ectopic pregnancy, and as a result, we could not have children. We applied for adoption while living in Sacramento, California, where I was participating in a joint USGS/US Army Corps of Engineers research project, and we were fortunate that the adoption agency found a five-week-old baby boy for us. We thought adopting a child would bring us back together. It did not. It only made matters worse.

Divorce is similar to death and resurrection. The old life ends, and a new one begins. But death is often preceded by a lingering illness. And even in this there are patterns: It has been said that, for someone on a spiritual path, soul evolution proceeds in seven-year cycles. The signs of the end of my old life began to emerge in Sacramento, seven years after my initiation into the practice of Kriya Yoga in 1960. They intensified 14 years later, in 1974, when we moved from Puerto Rico to Tampa, Florida. On the surface, my life appeared to be great. I had advanced rapidly in the UGSG, arriving in Florida as a GS-13 Supervisory Hydrologist at the head of the Technical Support Section, with 15 people under my direction. And we lived in a beautiful house in a great neighborhood. But we were both deeply unhappy.

We applied for adoption again, and were blessed with a beautiful little girl, but again, it didn't help. I submerged myself in work, traveling to Atlanta and Denver for training courses, teaching a meditation class one evening a week, and writing a book. Bobbi retreated into raising Pekingese dogs, African Violets and non-social drinking. It came to the point where about the only thing we had in common was sex. The breaking point came in 1976, just before my first book was published; my desire to escape was overwhelming. I resigned my job, went on a coast-to-coast book promotion tour, and never returned to our home again. It wasn't that I didn't love Bobbi and our adopted children, I just wasn't on the right path, and I knew that if I didn't do something about it, things would only get worse, and I felt that a mental breakdown was imminent.

When I gave my two-week notice, Dr. John Moore, my supervisor and friend, couldn't believe it. No one left government employment after 12 years, especially not when their career had been as successful as mine. He called my departure date my "self-destruct date." I gave up nearly everything I had accumulated. I signed everything over to Bobbi except for one of our three cars, which I kept, to go on the book tour. I left Tampa with nothing but a few personal belongings, a change of clothes and my severance pay. I did not mind giving up *things*, but the emotional results of leaving my marriage were devastating, for me and for my family.

I met Jacqui just after I had made the decision to leave my marriage and the USGS. We met when she came to the USGS office where I worked, awaiting my departure date, to pick up some maps for the consulting firm where she worked. She was very attractive, but I had said for years that I would never marry again. In fact, I remember saying on multiple occasions: "I won't do anything to end this marriage, but if I ever get out of it, I will *never* marry again!" The Universe, however, it seems, had other plans!

I learned later that Jacqui was very psychic as a young girl. She often knew what others were thinking and had visions of real events, including horrible things that happened to people, even violence and murders. Growing up in middle Tennessee, in a Christian family in the 60's, that was a problem. She had also had visions indicating that we would have a life together long before we met. I had no inkling of this, but there was a spark of recognition when we first met, something more than physical, but I often felt something like that when meeting someone new, and it usually just meant that we would be friends. We did, indeed become friends. One of the partners in the consulting firm she worked for had attended some of my meditation classes. Based on something she said about spiritual things, he suggested that she should meet me.

When she came to the USGS offices that day, Dorothy Cameron, the receptionist whose position was just outside my office door, was away from the office. Hearing someone coming in, and knowing that Dorothy was not there, I stepped out of my office and around the corner to see who was there. The pretty young lady standing there stared at me for a second and said:

"You're Ed Close!"

"Yes, I am" I replied. "Do I know you?

She said that she had been told that I taught a meditation class and wondered if she could attend. I was no longer teaching the class because I was planning to leave Tampa, but I was attending the meetings of the Self-Realization Fellowship Meditation Group in Tampa. I would not normally invite someone I didn't know to an SRF meditation. The proper procedure is for a person who is interested to apply for the SRF lessons to learn the basic meditation techniques, and the SRF Center Department would send them the address and phone number of the nearest group or meditation center. But, for some reason, I said; "Can you sit still without moving for an hour?"

"'If that's what I need to do, I'll do it!" She replied.

I could see the sincerity in her eyes, so I said: "I don't usually do this, but somehow, I think I'm supposed to."

I wrote: Thursday, 7:30 pm and the address on a piece of paper and handed it to her. When our hands accidentally touched, I saw flashes of red-orange robed monks in a monastery, somewhere high in the mountains of Nepal or Tibet!

I'd had some striking experiences in my life, but nothing quite like this. Because of this experience, when I went on my book tour, I invited Jacqui to meet me in Washington DC. We spent a wonderful weekend together in the Smokies, and we were married by a Justice of Peace on the day after my divorce was final, and two years later, on Friday July 13, 1979, we were married at SRF Lake Shrine in Malibu, California, in a formal Hindu fire ceremony officiated by an SRF monk. We have never doubted for one moment since, that we were meant for each other!

So, this is our story, because our lives have been wonderfully intertwined for more than 42 years as lovers, and as husband and wife this time alone, and for many more years into the distant past if our memories are to be believed.

Quantum Entangled Lives

The evidence suggests that our souls are eternal, part of God, and destined to return to Him again. But many lives may be lived during the thousands of years of descent from the peak of spiritual enlightenment, when nearly all souls are one with Cosmic Consciousness, to the lowest point of awareness in the cycles of conscious awareness, where very few are. It seemed that Jacqui and I had been together many times in the past. Memories of past lives may attest to this. Consider the following vignettes:

Ancient Egypt

The sun rose quietly over the *Aur* (the modern-day Nile River), and three sacred d'hauti, the arbitrators of time, moved slowly along the near bank, and a flight of cygnim swans wheeled gracefully reflecting the morning sun over the far side of the Aur. Standing on the stone balcony extending from our chambers, overlooking the river scene, I turned to Taya, B'nert-m'rewt (my Sweet Love) and said:

"The cygna remind me of the passage of time, my love. It grows colder in the north lands.

" Yes, Beloved. Does it remind you of the land of our fathers?" Taya smiled.

"Of course, it does, and I fear that the Truth that brought all this," I waved my hand around to indicate the beauty and luxury that surrounded us, "will fade with the descent of civilization into the cycle of darkness. I fear this grandeur will one day gather the dust of time and become lost to humanity."

"But, surely, we remember the majesty from whence we come, and will pass that memory on to our children?"

I sighed: "Yes, but we only vaguely recall now the time of our distant ancestors, remembering the coming of our progenitors from the stars, and their promise to return. I fear the descent into the Great Memory loss is still coming!"

"But, what of *Avram* , when he came to *Saqqara* a thousand years ago, and the connection with our *Mitannian* ancestors?" Taya queried, "Surely we have passed the nadir, and are ascending into the new age by now!"

"No, my love, I fear the worst is yet to come. The Nubians, Hittites and Philistines will eventually prevail against the civilization of Aten. We are but a remnant of the development of the last zenith of enlightenment - sadly, I fear, *the dying embers of civilization.*"

[*This is but one memory of many. Sometimes an individual conscious entity returns almost immediately after death of the bodily vehicle, sometimes there is a gap of many years, depending upon the desires and needs of the spiritually evolving atma (soul). But specific patterns of development can be discerned from life to life. I am only recounting life memories that have specific relationships to the mission of my present life.*]

The Greco-Roman Empire in Egypt

I climbed to the top of one of the porticoes at the entrance to the *Museion*, the structure that housed the library and museum of Alexandria, where I was privileged to be able to work and learn. My head was filled with thoughts about the teachings of Pythagoras. Plato and Empedokles. Across the Great Harbor, I could see Pharos, its light blazing across the night sky. I sat down, relaxed and enjoyed the cool breeze coming from the Sea across the harbor. I could see lights to the east, and hear faint sounds of music wafting on the breeze from the Jewish Quarter.

I was a child prodigy with thoughts and skills from memories of past lives, and in almost every one of them, my mother had been there. She was my wife in one, my older sister in another and always my closest friend, helper and protector. In this life, I had decided to become a scholar like my father. I had already learned Greek, Egyptian, Latin, Phoenician and Aramaic by the age of twelve years, and had begun studies of ancient tablets, scrolls and papyri in the Great Library. At the age of sixteen years, I began teaching philosophy, languages, mathematics, and natural science to help support my family.

My Father was a respected Greek scholar in this beautiful city at the western edge of the Nile delta, a city planned by Alexander the Great, and built by the Macedonian-Greek Roman General, Ptolemy. I was named Oregenes (son of Horus) by my mother, who was part Egyptian. My mother and father were, like many of the people of that time (100 to 200 AD), rather amazed by the adaptation of the religions of Egypt by the Ptolemaic Greco-Roman Pharaohs, to form a Royal Cult in an obvious ploy to absorb the citizens of Egypt into the Roman Empire.

There were groups of Jewish Gnostics and Christians in Alexandria, and my family had become members of the Christian branch of Judaism, which they found more attractive than the panoply of Egyptian and Roman gods. My father actively promoted the teachings of Jesus in his lectures, and, as a result, he died a Christian martyr. He was persecuted and then publicly executed for heresy by order of the Emperor Septimius Serverus when I was seventeen. I loved my father, and would have died with him, had my mother not intervened. On that terrible day when my father was arrested, she forbade me to leave the house, and even hid my clothes so that I could not go to my father's defense. I was very proud of him, because he did not renounce his convictions, including his belief in the resurrected Jesus, as many did, when confronted with the threat of torture and death.

Resting on the roof of the portico, I fell into a reverie, in a state between wakefulness and sleep. I heard a sound, and when I looked up, the resurrected form of Jesus stood before me, bathed in a blue-white light. He spoke to me, saying:

"You were raised up at this time for a purpose. You have a mission to fulfill: You must faithfully translate my words from the writings of my people."

I wanted to ask why *me*, but I was unable to speak. He continued:

"The Romans, and those who come after, will distort and falsify my words for their own ends. You will construct the link to the future that will renew my message in the end times of this cycle."

I began translating and studying the ancient Judean and Christian texts stored in the library in earnest after this experience. I knew this was important: it was to be my life's work.

I have other memories of the past. There was the life of a monk, for example, and Jacqui also remembers our lives together as monks, when we were both men. There was a time in India, when we were helping disseminate the teachings of the Indian and Tibetan Spiritual Masters. There is memory of a French lawyer-mathematician, Pierre de Fermat (1607–1665), and a German professor of mathematics and theology with an interest in linguistics and archaeology, Johann Liebknecht (1679–1749), a friend of the famous polymath, Gottfried Wilhelm Leibniz. The experiences of these lives have a direct relationship to, and effect on the mission started in Egypt as Amenhotep II and Oregenes. The mission, carried forward from 255 AD, has been to help bring certain specific spiritual, logical and mathematical aspects of the higher knowledge from the last peak of consciousness through the descent, into the current ascent that began about 500 AD.

I have written about some of these past-life memories in other publications, including my first book, *The Book of Atma,* Libra Publishers, New York, 1977, so I will not more spend time on them here, but instead, I want to stay focused on our current life together, and how the promise of Jacqui's precognitive dreams that I would take her to where she was supposed to be, and that she would help more people than she could imagine, would be fulfilled.

Jacqui's Wonderful Life of Service and Devotion

As I write this, it has been only 16 hours since I held my sweet love in my arms as she breathed her last breath and was carried away by Angels to the Kingdom of God. Just before this happened, I held her close, my arms wrapped around her, as she was racked with unbearable pain. She asked God *why He had not taken her home the night before*. She pleaded with Him, and we appealed to *Babaji*, the yogic incarnation working with Jesus. She asked him *when* she could come home, and he replied "as soon as possible." I held her close, our foreheads touching, and chanted "Om Babaji! Om Babaji! Om Babaji! Om! She breathed easier then, and a few more chants and breaths later, she was gone from her pain-racked body, and I had fulfilled my part in the promise that God had made to her all those years before: **I had taken her to where she was supposed to be, and she had helped more people than anyone would ever have imagined**, and now, she was home, in the presence of the Divine Beloved.

Thus ended the present life of a wonderful and beautiful woman, a real-life example for others; a life that went from abuse, despair, drug and alcohol addiction, to as near to sainthood as one can come in a single lifetime. But what about the journey? How did she help thousands, if not millions of people? I can only share the highlights here.

The Long Journey Home

After leaving Tampa, we spent some time in Missouri with my parents while considering moving to Rolla, one of the places where I had gone to graduate school, and where I began my career with the US Geological Survey. But the USGS had a hiring freeze in place at that time, so we decided to move to Denver, where some close friends lived. I took a job as handyman in the apartment complex where we lived in Lakewood, and Jacqui found a job in Golden, while I waited for the USGS to lift its hiring freeze. There would be an opening for a hydrogeologist in Denver, and I was well qualified for the position. But time passed, and the job didn't materialize. Jacqui did not enjoy the cold winters in Colorado, we had friends in LA, who had moved there from Florida, including Jacqui's best friend, Lana, and California was the biggest job market in the nation at the time, so we moved to LA.

The Authors circa 1979

In LA, we enjoyed being close to SRF temples, holy places and regular *satsang*, the company of friends who were on the same spiritual path. Jacqui had the strong sense of being where she was supposed to be, as did I, because being able to attend temple on a regular basis as well as SRF's annual World Convocation, helped us to strengthen our devotion to God and each other and we moved to the "propelled" state of realization, which would prove to be crucial for what was to come. Swami Sri Yukteswar, the guru of our yogic guru, paramahansa Yogananda, and thus our *paramguru*, described the stages of the human heart on the spiritual path in *The Holy Science*, pp 55 - 59:

> 23. There are five states of the human heart, viz., dark, propelled, steady, devoted, and clean. By these different states of the heart man is classified, and his evolutionary status determined. {The term "man" is generic in this context, and includes all races and genders.}

> 24. In the dark state of the heart man misconceives, i.e., he thinks that this gross material portion of the creation is the only real substance in existence and there is nothing besides. This, however, is contrary to the truth, ...and is nothing but the effect of Ignorance, *Avidya*. ... This state of man is called *Kali*; and whenever in any solar system, man generally remains in this state ...the whole of that system is said to be in *Kali-Yuga*, the dark cycle. {See Part X: Cycles of Consciousness and Time}

> 25-26. When man becomes a little enlightened, he ... begins to entertain doubts as to the substantial existence of [material creation]. His heart then becomes propelled to know the real nature of the universe and, struggling to clear his doubts, seeks for evidence to determine what is truth.... In this state [human beings], becoming anxious for real knowledge, need help of one another; hence mutual love, the principal thing for

getting salvation, appears in the heart. By the energetic tendency of this love man [the propelled soul] affectionately keeps company with those who destroy troubles, clear doubts, and afford peace to him; and hence avoids whatever produces the contrary result; he also studies scriptures of the divine personages scientifically. In this way, man becomes able to appreciate what true Faith is, and understands the real position of the divine personages, when he may be fortunate in securing the godlike company of some one of them who may kindly stand to him as his spiritual preceptor, *sat-guru*, the Savior.

It is necessary to say a little more about divine personages, spiritual preceptors, Sat-guru and the Savior because they have played such an important role in our lives. Wrong moves in life, like wrong moves in attempting to solve the cube, can lead to wanderings and entanglements that obscure the goal and result in years, and often whole lives, with little or no spiritual advancement. The loving guidance of an enlightened one, a divine personage, can help a propelled seeker avoid unnecessary entanglements, and thus avoid much suffering. The movement of the masses of human souls is very slow at the present time, with many losing their moral compasses completely, wandering aimlessly, losing hope, and believing there is no hope. This is the *Kali* state described by Sri Yukteswar. (See more about this in Part X: Cycles of Consciousness and Time.) In these times, we need the guidance of a true spiritual preceptor, representative of the Sat-guru or Savior.

When a soul embarks upon a spiritual path in these times, some guidance will be needed, and fully enlightened spiritual masters are few and far between. There are God-realized masters, however, in every time period, and many lives of wandering can be avoided by obtaining the guidance of one of them. Jesus and Krishna are perhaps the best-known divine beings who helped us through the most recent dark ages. But not all divine personages and spiritual preceptors are known to the world. Because of ignorance of the true nature of spiritual consciousness and attachment to physical forms, followers of each religious organization promote the idea that only their dogmas and their spiritual preceptor/savior is the only true way to God. But, why would a loving God allow untold numbers of souls inhabiting the bodies of aboriginal peoples live and die with no path to salvation because they never had the opportunity to hear of Jesus, Krishna, or Buddha? The truth is, all who attain Christ Consciousness are one. All God-realized masters are one. In Cosmic consciousness there is no plurality.

We both found good jobs in Southern California: Jacqui as administrative assistant to a vice-president of First Interstate Bank in downtown LA, and I as hydrologist and environmental engineer with Ralph M. Parsons Co. in Pasadena. But, even with good jobs, we could not scrape together enough money to buy a home, so when a high-paying opportunity as environmental planner became available at Yanbu Industrial City, which was being built at the west end of the Trans-Arabia pipeline with Saudi Arabia Parsons, I applied. It was an opportunity to save enough money in eighteen months to have a down-payment on a modest home.

Jacqui's precognitive dreams told her that I would take her where she was supposed to be, and when we moved to California, and lived near SRF temples and holy places, and many friends from this and past lives, she was happy. But she had not yet fulfilled the second part of the prediction, about helping more people than she ever dreamed possible, and she never dreamed that the next step would be to the Kingdom of Saudi Arabia!

Yanbu Industrial City

It was the first time since we were married that we would be apart for more than one day. I would have to enter the Kingdom of Saudi Arabia on a 90-day work visa while the residence visas were being processed. It took patience and persistence to obtain family-status residence visas. When I stepped off the plane in Jeddah, the gateway to the Holy Cities of Mecca and Medina in 1981, it was like landing on another planet in a distant star system. Everything about Jeddah was radically different from any city in Western civilization. The smells of charcoal, incense and spices wafted on the hot night air, with an unpleasant undercurrent odor that I couldn't identify.

There were colorful neon signs in complex sweeping Arabic script, but no English translations. I determined then and there, that, if I was going to live in Saudi Arabia for eighteen months or longer, I must learn Arabic. My native tongue was *Ozarkese*, an American-English dialect, but I had learned English grammar, German and some Spanish by the age of twelve, so I had an ability to mimic sounds and learn new idioms fairly quickly. But Arabic was a very different language than the Teutonic and Latin derivatives I knew. Arabic script is read from right to left, root words are tri-consonant, and can be sounded without knowing what they actually meant because the script is almost completely phonetic, a feature English all but lost in the transition from Low German to Old English, and the inclusion of Latin and Greek words and verb forms.

Being on the other side of the world (eleven hours ahead of California) I couldn't sleep, even though it was about 2 AM when I finally got to my hotel room. Saleem, the Company man who met me at the airport and helped me through customs, told me several times that I had to be up and on the bus to

Yanbu at 7:00 AM. Since I couldn't sleep, I began reading an English translation of the Koran. I wanted to learn as much as I could about the Arabic Islamic culture while working in the Kingdom.

I have written more in detail about my experiences in Saudi Arabia in my autobiography, which is currently an incomplete manuscript to be published at some point in the future, and in the first part of this book I wrote about learning to solve the cube while in Saudi Arabia. But the purpose of this part of the book is to focus on patterns of life that exist like patterns in the process of solving the cube, so I'll focus on the aspects of this life that directly involve Jacqui and me, and our work together toward the goal of the mission started with Amenhotep II in Egypt and continued in Alexandria with Oregenes Adamantius around 500 AD. That goal is to preserve the truths suppressed and subverted by the later dynasty pharaohs and the Roman Emperor Justinian I, concerning the spiritual nature of reality, into the ascending yugas. I will therefore skip over many irrelevant details.

I think it is pertinent to briefly mention a weekend excursion into the Hijazi Mountains with two other Yanbu ex-patriot employees during the 90-day period while the residence visas for my family (Jacqui and Joshua) were being processed. As I mentioned, above, I had my first Rubik's cube with me, and passed some of my evenings learning to solve it. But the six-day, ten-hour day work week at Yanbu Industrial City, really just a construction camp at that time, in a desolate location on the Red Sea, 300 kilometers north of Jeddah, was exhausting, and we soon ran out of anything to do on-site on the one-day weekends. One weekend, Fellow American, Giorgio, a landscape architect and Douglas, an engineering architect from the UK and I decided to escape the boredom of the construction site.

Giorgio had a jeep, so we decided to drive up to the most prominent mountain visible from our location, a bare rocky peak about 45 km north east of our location, and explore. I didn't remember hearing in my arrival briefing, or reading in the fine print of my work visa, that ex-pat employees were not allowed to go beyond the boundaries of the construction site without a written permit signed by the Saudi Director General of the Royal Commission and the local Saudi Prince. I found out later that it took days, or even weeks to get permission to leave the construction site, the reason for leaving had to be a legitimate reason, approved by the company administration. The desire to escape the boredom and explore, would not be an acceptable reason. If Giorgio and Doug knew this, they didn't bother to tell me, and decided that the prospect of being caught was an acceptable risk.

We drove to the ancient fishing village of Yanbu Al-Bahr on the Red Sea about 20 km north of Yanbu Madinat Al-Sinaiyah (Yanbu Industrial City - our construction site), then turned inland to an even smaller village, Yanbu Al-Nakl (Yanbu in the Hills), and thence on to a wadi (dry wash) where the remains of an ancient earthen dam suggested that the climate here had been more moderate and humid sometime in the past. We drove up the wadi as far as the jeep could go, and backpacked on through a siq, (narrow gorge) in dark bluish igneous rock with bands of white quartzite, leading to a steep ascent up the south face of the mountain. About halfway up, we pitched a temporary camp. Douglas, the quintessential Englishman, brewed tea, and we had tea and biscuits (hard cookies Doug had received from home) followed by oranges the Royal Commission had shipped from Lebanon to the construction site periodically.

Doug and Giorgio relaxed, and I climbed to a prominent ledge some fifty feet higher up the mountain side and found a spot to sit and meditate. Off in the distance to the southeast, I could see the spires and minarets of a sizable city, that I learned later, was the Holy City of Medina, one of the two Holiest cities in Saudi Arabia, that every devout Muslim wanted to visit on spiritual Hadj (pilgrimage) at least once in his lifetime.

As the sun began to descend in the west, we realized that we would have to hurry to get to the jeep, out of the Wadi and to the road before nightfall. It would be easy to get lost in the dark and have to spend the night here and risk not being back to work in the morning. We made our way back to the jeep, but before we could reach the wadi mouth and the road, a group of five or six Bedouins on horseback appeared to be advancing from the east to intercept us. Giorgio shifted gears, and turned west toward a side of the widening wadi that looked negotiable. The little jeep successfully climbed the slope, and bounced on the cobbles of the pervasive desert "pavement" on the other side. The horsemen gave pursuit, but we out-ran them and when we got within sight of Yanbu Al-Bahr, they gave up and turned back.

Holding on for dear life as we bounced across the desert, I didn't understand why Giorgio seemed so desperate to avoid being overtaken by the Bedouin horsemen. He spoke some Arabic, and had had no problem communicating with the townspeople in Yanbu Al-Nakl, on the trip out. I learned much later, that the mountain we had climbed, was Jebel Al Radwah, one of the holy places where Mohamed went to pray. Technically, as non-Muslims, we could have been beheaded as infidels trespassing on Holy Ground. At best, they would have arrested us and taken us to the local Sheik, who would have put us in jail and notified the Yanbu Al-Sinaiyah administration, and we would have been expelled from the Kingdom. At worst, we would have simply disappeared; beheaded and buried in the desert.

Even though Jacqui was not directly involved in this misadventure, I have included a brief recounting of it because it is just one of several occasions during this lifetime in which the outcome could have been disastrous, resulting in my death and preventing us from accomplishing what we set out to do together in this life. During every such life-threatening event, I felt confident that I would survive. I always sensed the presence of Babaji, one of our spiritual mentors protecting me.

A Tense Moment on the Jeddah-Yanbu Highway

A few months later, when the residence visas were approved, I was still blissfully unaware of the requirement to get my passport, which was in the company's admin office, and a regional travel permit before leaving the immediate area of Yanbu Al-Sinaiyah and Yanbu Al-Bahr. I had received messages from the Company back in Pasadena and from Jacqui telling me that they had received the residence visas, and the Lufthansa flight Jacqui and Josh would be on and their ETA in Jeddah.

The bus ran between Yanbu and Jeddah only once a day, Sunday through Thursday. It arrived in Yanbu from Jeddah about 11:00 am, unloaded, and immediately turned around and headed back to Jeddah. I had to be sure to be there to board the bus for the return trip to Jeddah, in order to check in to the Arabian Sands Hotel, contact Saleem and be at the airport to meet the incoming flight about midnight, wait for Jacqui and Josh to get through customs, take them to the Hotel, and get on the bus back to Yanbu at 7:00 AM the next Morning. I was so eager to see them, that it didn't occur to me that I might need to go to admin to get my passport and visa, and no one had yet bothered to tell me about the requirement. I proceeded the same way I would have back in the US, and simply jumped on the bus and settled down for the three-hour trip to Jeddah.

The trip to Jeddah was uneventful and I had time to buy some welcome-to-Saudi Arabia gifts for Jacqui and Josh, have dinner and relax before Saleem came to drive me to the airport. It seemed to take forever for them to get through customs, even with Saleem's help. Jacqui had brought thirteen pieces of luggage, including suitcases, boxes and even a couple of trunks. Of course there were no cell phones in 1981, and calling from Yanbu to California involved a long wait in lines at the Company's Communication Center, the connection was poor, fraught with disconnects and very expensive. When we had talked or communicated by air mail, Jacqui always asked to me find out from ex-pat wives what necessities were in poor supply. As a result, she brought a year's supply of several things, including toilet paper and brown rice. All thirteen pieces of luggage and Jacqui's carry-ons had to be thoroughly searched by the customs agents for things forbidden in the Kingdom: alcohol, weapons, pork products, Christian Bibles or tracts, and pornography, which included any magazine, book or other printed material with pictures of partially-clad women, like, e.g., mail-order catalogues.

It was very late by the time we got back to the Sands Hotel, but Joshua was wide awake, as was Jacqui, so we talked until daybreak, had a light breakfast in the hotel restaurant, and helped Saleem get all of the luggage onto the bus. I expected an uneventful bus ride back to Yanbu, but that was not to be the case. About half-way between Jeddah and Yanbu, there is a small village called Rabigh. The bus driver regularly stopped there because there was a gasoline station with a restroom and snacks.

Passengers could get out and stretch their legs for a few minutes. But as we were approaching the village, suddenly there was a road block, manned by several uniformed guards of the Saudi military. A cache of weapons had been discovered somewhere in the region, presumably for the purpose of a planned attempt at a coup to overthrow the Saudi Royal family. No one except the military was allowed to possess weapons in the Kingdom. One of the soldiers came onto the bus waving a machine gun, demanding to see the passports of all passengers.

The soldier who came onto the bus carrying a machine gun, looked to be about fifteen years old, and he spoke no English. Jacqui was visibly shaken, especially when I told her that I had not brought my passport and igamma (work permit) with me. She was faced with the alarming possibility that they might drag me off the bus, and leave her and our four-year old son alone in a strange land far from home. She knew that, in Saudi Arabia, a woman was not allowed to drive or travel alone. She must be accompanied by her husband, or an older male family member like her father or an uncle, if unmarried.

When the soldier came to me, and I could not produce a passport or papers of any kind, he became agitated, waving the machine gun in my face, repeating *"Shoof! Al-bazboort, Al-bazboort!* The bus driver, a Somali national, tried to interceed, telling him in Arabic that I was an employee of the Royal Commission of Yanbu and Jubail, but it didn't seem to help. Just when it seemed that I was going to be forced off the bus, another passenger, a Somali man and his two wives, began shouting at the soldier. My understanding of Arabic at that time was very rudimentary, but somehow, I knew that they were telling him that I was an important scientist working for the Saudi Royal Commission, and that they would be in big trouble if they separated me from my wife and child. The soldier turned and glared at me and said:

"*Maa Angleezi?*" (Are you English?)

I knew that the Saudis were very unhappy with the British at that time because of the release of the movie, *Death of a Princess*, a dramatized documentary film produced in England, about the recent public beheading of Princess Misha'al and her 19-year-old lover for adultery. The British ambassador to Riyadh, Sir Albert J.M. Craig, had just been expelled from the Kingdom because of the film. I replied:

"*La, la, ani Amreeki.*" (No, I am American.)

He calmed down, glared at the driver, took one last look around the bus, withdrew, motioned for the road block to be removed, and waved the bus on.

Helping More People Than She Ever Dreamed Possible

Jacqui had survived lightning strikes and the death sentence of cancer, but in 1994, her health was still far from perfect. She had sciatica, occasional internal bleeding from cystitis caused by the linear radiation used to treat the cancer, frequent colds and flu, migraine headaches and sinus infections. Relief came from an unexpected source: While in St. Louis on business, I had found an interesting little shop called "Cheryl's Herbs".

One weekend, when Jacqui was suffering with a sinus infection and headache, I suggested we get out of the house and drive to St. Louis, perhaps dine at a nice restaurant and see a movie. She agreed, but as we drove (more than 100 miles by interstate highway), she felt progressively worse. So I decided to take her to straight to the herb shop, where they had all sorts of remedies. Maybe something they had would help. But the shop was in an out-of-the-way place in an industrial park in Maryland Heights, and I had some trouble finding it. Jacqui was not happy that I was apparently lost, driving around, back and forth between warehouses. When I finally found the shop, she didn't even want to go in; she felt miserable, and wanted to go home. I managed to talk her out of the car and into the shop.

When we entered the herbal store, we noticed a pleasant aroma, something subtly spice-like with a floral undertone, like gardenia or jasmine. Within a few minutes, Jacqui perked up and said: "My headache is gone!!! What is that aroma?" It was coming from a nebulizer running at the back of the store. A young man, a university student working in the shop as a clerk, informed us that they were diffusing "essential oils".

"What are essential oils?" I asked. Jacqui was a step ahead of me: "It smells like a blend of two or three fragrances; what is in it?"

"Oh, I don't know, we just put different oils in it, sometimes rose, sometimes patchouli ... whatever we like."

"But it cleared up my sinuses! My headache is gone! I want to buy some" Jacqui exclaimed.

The clerk pointed to a row of little bottles, and handed her a little wooden stick. There was one of these sticks beside each bottle.

"Just stick this into the top of the bottle to get a drop of oil to smell and see which one you like."

We ended up buying over $100.00 worth of oils and a diffuser. That was the beginning of a wonderful journey. Jacqui began an intensive study of the alternative health modality using essential oils, and **this was the beginning of a sequence of events that enabled Jacqui to help more people than she ever dreamed possible.**

The Healing Power of Essential Oils

In these days, when health care is conflated with insurance policies, and allopathic medicine has fostered an ever-expanding, multi-billion-dollar pharmaceutical industry that churns out hundreds of synthetic drugs with horrifying side effects daily, the modern world has forgotten that real medicines are found in nature in a variety of medicinal plants. Herbs and essential oils distilled from plants were the medicines of humanity in the ancient civilizations branching out from the fertile crescent of Mesopotamia, and the Garden of Eden.

Pharmaceutical drugs are created today by studying medicinal plants, identifying disease-resistant organic compounds that exist in the essential oils that circulate through them, carrying oxygen and nutrients to their leaves. The compounds that are identified are then synthesized in the laboratory, patented, and sold for great monetary gain. These synthetic drugs lack the complimentary organic compounds that exist in the source medicinal plants, and without them, the synthesized medicines have serious side effects. The relative strengths and composition of the natural organic compounds in medicinal plants vary from season to season, and with the composition of the soils in which they grow. Because of this, they can't be standardized, patented and controlled by the pharmaceutical companies.

Jacqui studied the chemistry of essential oils with Dr. David Stewart, the founder of the Center for Aromatherapy Research and Education (CARE). She took and passed the exam to become a registered aromatherapist with the National Association of Holistic Aromatherapy (NAHA) and was one of the first fully certified CARE instructors.

An Important Discovery

In a book we authored together, Jacqui wrote about a discovery we made that led to *helping more people than we ever dreamed possible*. The following are excerpts from that book, *Nature's Mold Rx, the Non-Toxic Solution to Toxic Mold*:

> In 1992 we returned to Ed's home state of Missouri to be near his aging mother. When the firm Ed was working for told him he would have to move to Chicago or Kansas City, Ed and Jacqui decided it was time to stop moving and open their own firm. So, in 1995, with two clients, they opened Close Environmental Consultants. Ed already had thirty years' experience in the environmental industry, a PhD in environmental engineering, and had been an advisor to more than ten Fortune 500 companies. Jacqui had grown up in family-owned businesses. Her business and accounting background proved most helpful in their engineering business.

Ten years later…

It was a cool, late October day in 2005, when a long-standing client of Close Environmental Consultants contacted us and asked that we do third-party sampling for mold in an apartment complex their company had purchased and was renovating. The apartments had been flooded, then evacuated, and later put up for sale. Closed and vacant for some time, mold had run rampant in the buildings. Our client was renovating the apartments, and his lender required that the apartments pass a mold inspection.

Mold and allergies are a big problem in this part of the world, but most people just accept it as part of life along the Mississippi River. When Ed went to the apartments to do the sampling, most of the water-damaged, mold-infested material had already been removed, and a cleaning company had already gone through the apartments and cleaned everything with a solution that they told the new owners would kill any mold. The sampling results showed that numerous molds, and especially molds of concern, which are commonly known as toxic molds, were still extraordinarily high when compared with mold levels found in the outside air, and also to levels typically found in the surrounding region…

The cleaning company had used a chemical their supplier told them would kill mold and anything else that was harmful to human health. They said: "It's the same product used for hospital clean rooms."

Based on the results of sampling, the product had not killed the mold. Or, if it had, the mold had already re-established itself... Reviewing the documentation and the Material Safety Data Sheets (MSDS) for the product the cleaning company had used, Ed learned they were bactericides, and suggested they acquire an industrial-strength fungicide and re-treat the buildings. They did a second round of tests that showed that the levels of mold were still extraordinarily high. [The client was perplexed. He had already spent a lot of money, and he could not finish the job and rent the apartments if they didn't pass mold inspection.]

Ed was discussing this situation with Jacqui (who had discovered essential oils in 1995 and became a registered aromatherapist in 2001), and she asked:

"What if there was a non-toxic solution to toxic mold?"

Jacqui urged Ed to talk to the client about allowing him to do tests in several of the apartments to see if essential oils would work. Essential oils are known to be anti-fungal...

Ed did not feel comfortable suggesting to his clients that they use essential oils to treat toxic mold infestations in buildings... Ed doubted it would work, and did not want to suggest anything without sufficient scientific basis ... He said:

"If I tell my engineering clients that I'm going to solve their mold infestation problem with aromatherapy, they'll laugh at me!"

Then don't call it 'aromatherapy'! Jacqui replied.

Lacking any other viable and economically feasible alternative, Ed decided to approach his client ...

Long story short... it worked! Not only did the before-and-after sampling show remarkable results, there was a residual effect. Since the oils used were not toxic to humans and pets, and even beneficial, home and business owners did not have to vacate the buildings being treated and wait for toxic chemicals to dissipate. The residual effect was an unexpected plus. In several test cases, mold levels remained far below outside levels for weeks, and, in some cases, even months after treatment. When the news of this success reached the owners of the company producing the oil blend that was most effective, we were invited to do presentations for distributors. And I was an invited guest speaker at the company's Grand Convention in 2006, a week-long affair with about 3,000 distributors in attendance. This was the beginning. In 2007, we published the book. We also produced a DVD of one of my first presentations, and a CD entitled "Essential Oils vs Mold in the body, which was mostly Jacqui's research on using essential oils to combat the health effects of serious mold exposure.

We became widely known and sought after for local and regional conferences, where we taught classes and sold our book, DVD and CD. This sent ripples through out the alternative health-care communities, bolstered the essential oil industry and raised the ire of the pharmaceutical industry. Even though the pharma lobbyists and activists within the government pushed the FDA to shut down claims of positive health benefits with essential oils, the flood of testimonials could not be stopped. We received frequent requests to speak at conferences and hundreds of requests for consultations. Here's a sample of the hundreds of the positive feedback we received:

"**Nature's Mold Rx more than just a book**, December 2011
This book, or the DVD can save your life or the life of friends and family. Mold is a silent killer, it is up to us to take care of ourselves, this book will help you do that. It tells you everything you need to know about the mold that is in all of our lives, ... We have seen the results of the applications the book teaches, and they work, There is so much information in this book. It is a treasure, a guide every home should have." – JK, Missouri

"**Very informative video**, January 2012
I own this video and appreciate the scientific approach Dr. Close brings to the table....I have used the product he suggests and know that my home is free of mold because of this and my family is much healthier." – EL, Minnesota"

"**Five stars to Nature's Mold Rx**, September 2012
Excellent book.
Information everyone should know, and explained in detail that is easy for everyone to understand and interpret". – Bella

"**Good book**, March 2013
Opened my eyes and got me to understand what we were going through. Not easy to get the companies to work with you in dealing with it naturally, but in the end our health was the biggest concern and we learned a ton from this book". - SM, Florida

"Helped us win the battle, August 2011
Last year, we who have always been heathy started getting sick and finally discovered we had mold in our home. Toxic mold ... it was EVERYWHERE by the time we realized it.... My children developed rashes and fevers and flus and allergies. My husband suffered recurring sinus infections. My lips and mouth would swell and itch.... The information in this book and the mold dog were the answers to our desperate, desparate prayers and were what finally turned the tide in our battle against mold". – JM, California

"5.0 out of 5 stars Nature's Mold Rx the Non-Toxic Solution to Toxic Mold, May 2009
Due to a recent toxic mold exposure where I work, I have spent months researching extensively in regard to toxic mold and mold remediation. This book offers hope to the 25% of the population adversely affected by mold exposure. Toxic mold survivors face the untold heartache of losing property, health, jobs, or worse yet, many times a lifetime of disabilities that will eventually end in a terminal illness. The sooner this product and the patent pending protocol outlined in this book are considered standard tools in every mold prevention and remediation project, the more lives, property and millions of dollars in liability issues will be saved.... I can personally attest to the fact that the methods outlined in this book work." – PL, USA

"I purchased this book because one of my best friends died from having toxic mold in her lungs. The doctors had no treatment for her... If I had learned about the information presented in this book while she was still alive, I am certain that she would have been able to recover. ... I use this book very often to share with others the power of essential oils. I also share it with them so that they may realize that some of their symptoms may be a result of toxic mold exposure ... I would recommend this book to anyone who has water damage in their home." DC, Missouri

"I have been sick from a bad exposure to mold and done a lot of research on the subject since most doctors don't hardly acknowledge it as a problem. Dr. Close has found the answer to mold elimination. It is not toxic either which is perfect since I'm already suffering from a reaction to a toxic substance. I can take the diffuser with [the] oil wherever I go. I had all my belongings stored in a garage that ended up turning everything into mold. The [essential oil cleaner] removed all the surface mold and the diffuser did the rest. I owe a debt of gratitude to Dr. Close and Jacqui for all the work they have done to help all of us who have suffered form mold exposure.... I wish they would diffuse all our schools as a precaution to illness from mold to our children. My life will never be the same although I feel like I at least have a chance now that I know what to do to keep my environment safe." – SH

These testimonials mention me primarily, but they are really testimonials to Jacqui's love and desire to help those in need. None of them would have been written by those whose lives were saved and improved, without the heart, support and substantial input of the tireless, selfless woman who was my partner in everything I did: *my soulmate Jacqui*. She is the one who suggested the possibility that essential oils could be the non-toxic solution to toxic mold. She is the one who inspired me to try to find an engineering solution, a way to apply essential oils effectively and set up the experimental framework to document the results obtained by independent EPA approved laboratory analysis.

It was Jacqui's heart-felt desire to help others that made all this possible. And it was through this application that the prediction delivered to her by spiritual guides, - angels, if you will- many years ago, before she ever met me in this life, that she would help more people than she ever dreamed possible.

Important Networks of Friends

Through our association with Young-Living essential oils for more than 15 years, Jacqui and I met thousands of people, as we spoke at Young Living world conventions, regional conventions and local gatherings about how people could use natural essential oils to take control of their lives. We spoke about the discovery of the effectiveness of essential oil blends to eliminate mold infestations, and Jacqui counseled people whose lives had been seriously impacted by mold exposure. For several years, she hosted monthly on-line webinars on the many uses of essential oils. We became well-known as guest speakers for Young Living and travelled across the country and as far as Australia. I also spoke about quantum physics and consciousness at a number of conferences after the publication of my third book, *Transcendental Physics,* in 1997.

When I qualified for membership in the International Society for Philosophical Enquiry (ISPE) in 2008 and connected with Dr. Vernon Neppe, Jacqui and I already had a world-wide network of Self-Realization Fellowship friends, Young Living friends, and others interested in subjects basically ignored by mainstream science. But my association with Dr. Neppe opened the door for us to connect with another network, a number of professionals who were working outside the box of mainstream materialism. Dr. Neppe was, and still is, a member of the Advisory Council of The Academy for Spiritual and Consciousness Studies (ASCSI). I was invited to speak at two of their conventions and Dr. Neppe and I presented a workshop on TDVP in 2016 in Chapel Hill North Carolina.

Subsequently, I reconnected with the Rhine Research Center, the current organization carrying on the research started by Dr. J.B. Rhine and his wife Louisa, on the Duke University campus in 1935 as the Duke Parapsychology Laboratory. As mentioned earlier, David Stewart and I had performed some controlled experiments in parapsychology and communicated with Dr. Rhine in the mid-1950s. Valuable contacts were made through these organizations that led to my becoming one of the early members and supporters of the Academy for the Advancement of Postmaterialist Sciences (AAPS), and communicating with some of the leaders in the field of consciousness studies.

During the period 2000 – 2018, Jacqui was able to help many people to take control of their lives and survive, as she had, after conventional medicine had given up on them. Between helping people deal with toxic mold, cancer, chronic pain, and other challenges, Jacqui literally helped more people than she ever dreamed possible.

Jacqui's Great Ordeal and Revelations

As a devout Christian and SRF member, Jacqui often consulted with spiritual counselors. We attended SRF convocations every year that we could for many years, and one year, in a session with an SRF nun, she confided that, while she was living a wonderfully blessed life, she had nearly died on six separate occasions, and suffered with a lot of pain. "Why", she wondered, "do I have to suffer so much?" The nun answered: "Don't you know, my dear, that you are finishing the karma of several lives, all at once?"

Much of the material in later part of this section is based on notes written by Jacqui while she was in the hospital in an altered state of consciousness. She jotted notes on napkins and scraps of paper, until I brought her a blank notebook. Some of it was written in a strange alpha-numeric code that I could only partly decipher, and even she couldn't decode all of it herself, even after she got home. Some of what God revealed to her during her NDE experiences she was not at liberty, or willing, to reveal. I will strictly adhere to her wishes in this regard. She started re-writing much of the material upon returning to a normal state of mind, shortly after being discharged from the hospital in late August 2018, when it looked like she would recover, more-or-less to what had been her normal life before the events of July 13, 2018 that resulted in the first of several trips to the emergency room. That meant continuing with peritoneal dialysis, which had allowed her to lead a restricted, but relatively normal functional life for six years after her acute kidney failure in 2012.

More Troubles and Travails

I have avoided mentioning Jacqui's brush with death via acute kidney failure until now because it did not impact her ability to help others very much. She had the amazing ability to rise above her health problems and function as a bright, thoughtful, and cheerful angel, when focused on helping others, even during the last year when she was in severe pain almost all the time.

In 2011, after Dr. Vernon Neppe and I had published the first edition of *Reality Begins with Consciousness*, I submitted a paper on TDVP for the 2012 'Toward a Science of Consciousness' Conference in Tucson, and it was accepted as a poster presentation. We had a lot of potentially very good things going on. I had been part of the Young-Living Frankincense Trail Documentary in Egypt and Jordan in 2010; I was rapidly advancing in rank in the International Society for Philosophical Enquiry; we had been traveling and presenting at conferences for Young-Living Essential Oils; I had also been invited to speak at a Pythagorean Conference in Louisville Kentucky in November 2011; our book and DVD on the solution to toxic mold were selling well, we had been invited to speak at Young-Living Regional Conference in Brisbane Australia, and the future was looking very bright.

While at the Toward a Science of Consciousness Conference, we decided to move to Tucson. We had friends who lived there, I knew people at the university, and while I was busy at the conference, Jacqui found us an apartment in North Tucson, not far from the conference site at the hotel in Ventana Canyon. After the conference ended, we proceeded to put our house in Missouri on the market, and begin our move to Tucson, with plans to expand our environmental consulting and Young-Living essential oils businesses there. But 2012 turned out to be a disastrous year for Jacqui and me. The conference in Kentucky was cancelled after we had spent a lot of money preparing for it, a tree fell on our house in Missouri during a storm, and when we returned from the trip to Australia, Jacqui suffered acute kidney failure and wound up in a hospital in Tucson, near death for eleven days. Jacqui would have to be on dialysis for the rest of her life.

As I mentioned in the section titled "Opportunity and Tragedy" in Part V, Jacqui had survived cancer in 1985, when the chief oncologist at Massachusetts General Hospital in Boston gave her only two months to live, but as a result, we had no insurance, and Jacqui was not old enough to start drawing social security. The staggering medical expenses in Tucson and the prospect of the on-going expenses of dialysis for the foreseeable future and the impact on our businesses, made it necessary for us to abandon the plans we had for a new life in Tucson. Fortunately, our house in Missouri hadn't sold. With the help of friends, including Jacqui's best friend Lana, who drove from LA to help when Jacqui got out of the hospital, and my best friend David, who loaned us money to finance our move back to Missouri, and Young Living friends Robin and Shaun, who helped me pack up our belongings and get our furniture loaded into a U-haul van, we limped back home to Missouri.

An accident in New Mexico, and a dialysis center in Texas that couldn't take Jacqui when we arrived there, even though we had made reservations for treatment in Texas and Arkansas prior to leaving Tucson, made the trip a very stressful one, and after we were resettled in Southeast Missouri, and I came back from a trip to Puebla Mexico where Dr. Neppe and I were guest speakers at the *Congresso de la Sciencia y Spiritualidad*, I was hospitalized for four days.

These blows to our health and finances were staggering, but we survived, and Jacqui's indomitable spirit would not surrender. Even with the limitations of the necessary routine of daily peritoneal dialysis, we continued to attend the world Convention in Salt Lake City, and the SRF Convocation in LA for a few years after our return to Missouri. Even though She could not function normally for more than four or five hours per day, Jacqui held weekly on-line seminars and we travelled around Missouri, and to Kentucky and Minnesota to do Young-Living presentations.

I continued working with Dr. Neppe, on our research, writing and publishing. In addition, I continued doing some environmental consulting for several years, to help pay bills. I did some Young-Living conference presentations by myself when Jacqui just couldn't go, but we managed to go to science and spirituality conference in Chapel Hill North Carolina and Scottsdale Arizona together. By 2017, traveling was just too much for Jacqui, and when I was invited to become part of the Academy for the Advancement of Postmaterialist Sciences, I had to go to the first AAPS meeting at Canyon Ranch near Tucson, alone. And I had to go because I knew it was an important event for all concerned.

Another Ordeal

We knew that Jacqui's remaining time on Earth was limited; but we didn't want to accept the inevitable. For a time, her progress after her renal failure in 2012, was phenomenal. Dr. Frank Braxton, her nephrologist, said she was the first dialysis patient in his long career for whom he had approved actually reducing the number of exchanges per day. She was asked to speak to groups of nurses and patients about the advantages of at-home nightly peritoneal dialysis over out-patient hemodialysis in hospital every other day. Much of her improvement and sustained vitality, however, was due to her freedom at home to use essential oils in addition to the allopathic treatment of nightly dialysis.

Even with the advantages of our knowledge of alternative health modalities, and Jacqui's strong will, we were both greatly relieved when our son Joshua came home to help me when care for her became a full-time, 24-hour job. I was very happy to do everything I could to make life as normal, enjoyable and peaceful for her as I could. She deserved to be treated as the beautiful princess-soul that she was, and I was thankful to God that my health was exceptionally good, especially for someone my age, so that I could do most of the care for her as long as I could. Without the blessing of our son's return, care for my love would surely have become completely overwhelming for me. As Jacqui became unable to walk, and needed help to get into and out of bed, go to the bathroom and bathe, we divided the chores between us: Josh did most of the cooking, shopping, driving and bill paying, while I devoted myself to attending to Jacqui's more personal needs.

The beginning of the end started on the night before our wedding anniversary. We knew something was terribly wrong when the well-established routine was upset by Jacqui becoming unable to do her part, which consisted of hygienically connecting her catheter to the dialysis machine after we brought the dialysate bags into the bedroom, set up the cycler, and closed the door. When I heard her "okay", I would go in, make sure everything was in order and help her get comfortable for the night. This night, time went by, and we heard no "okay". I tapped on the door, and asked if everything was OK. When there was no reply, I went in and found Jacqui standing in front of the mirror, repeating a series of words and numbers over and over. I put my hand on her shoulder and said:

"Jacqui, Sweetie, you need to get on the cycler."

She said "Okay", so I went back out of the room to attend to something else. When I came back about 15 minutes later, she was standing in front of the mirror again, reciting the same sequence of words and numbers. I called to Josh, and told him what was happening. He came in and we managed to pull her away from the mirror and tried to get her connected to the machine. We did something wrong, and the machine began to alarm. As we tried to figure out what we needed to do, and couldn't, we began to panic, and I called the PD nurse on the emergency line. When I told her what was going on, she said "You need to take her to the emergency room right away!" By the time we got her out of the house, into the van, and drove to the ER, it was after midnight, it was Friday July thirteenth, 2018, and our wedding anniversary.

It turned out that even though she was observing a strict diet, with restricted fluid intake, and doing three dialysate exchanges nightly, her electrolytes had gotten out of balance, and outside of acceptable ranges somehow, resulting in her brain being flooded with ammonia, a toxic gas. Excess ammonia on the brain is known to impair memory and brain function, shorten attention span, cause sleep-wake inversions, edema, intracranial hypertension, seizures, ataxia, and coma.

In the hospital, they were able to get the parameters back within bounds and she was discharged to come home. But two days later, it happened again, and this time it was worse. She was incoherent, refused to get on the dialysis cycler and wouldn't allow us to take her out of the bathroom. After struggling until we were exhausted, we had to call the PD nurse and 911. I'll not describe the travails of the next one-and-one-half months here, I'll just insert what she wrote about the ordeal herself, after she was released from the hospital on August 22nd, 2018.

The next section describes Jacqui's experiences from July 13 to August 23rd, in her own words. There are some parts that appear to be repetitive, but they were written at different times, so

I have not edited them because I want to honor her wishes to let her say what she wanted to say, in the way she wanted to say it.

Jacqui's Hospital Ordeal

> *"The reality of my life cannot die, for I am indestructible consciousness."*
>
> **Paramahansa Yogananda, Metaphysical Meditations**. Page 41, 1994 Printing

My Story: After Four (4) NDEs in My Life, I Have Five (5) More in One Month

By Jacquelyn A Close, Copyright, August 28, 2018

My introduction to Our Readers:

Wondering what this is all about? Well, I am writing this for friends and family who would like to know more, and understand what happened to me during July and August of this year. Some of you reading this have already seen a rough draft of this, and your thoughtful comments have informed and impacted the content of what you are about to Read.

This will be the longest version I've written. It will contain a lot of the details of my hospital stay and its effects on me. This version will contain all the information I am able to share. I am not editing for length at this time, I am simply trying to get my story out. Your comments and questions will help inform edits that will be made before posting parts of this on Ed's blog, www.ERCloseTPhysics.com, and later appropriate portions will be included in one or more books that Ed and I will write together, or separately. Your input is extremely important to us.

I am writing this for the purpose of helping you understand both sides of the experience, including diagnosis and prognosis, as well as my understanding of what was happening from my perspective. My great hope is that you will send your written comments and questions, both positive and negative back to me, or Ed, in case I am no longer in this miserable body, which will help us edit and adjust the text appropriately.

Ed and I are already working on a new book together, called **"Quantum Entangled Lives, A Cosmic Love Story."** Some of the material you read here may be included in that book. I want to write a separate book about this experience, called, **"Wacky Jacqui and Her Trip with God Through La-La Land, The Spiritual Aspirant's Journey on Earth."** This will be included to a large degree in the book I write. So again, your comments, questions, your contribution for the betterment of this work is most sincerely appreciated, and we may wish to quote some of you in the book or on the covers of our books. That is why, for legal purposes and for keeping my life as simple as possible, I had to require that you send me an email requesting to be on the list to receive this information. The story is taken from detailed notes I made during the experiences (most in a short-hand code I invented for the purpose). Here it is, after some editing and review for correctness.

My Story of July and September 2018

I have been through five (5) Near Death Experiences (NDEs) since July 13th, 2018. And while there are many things that I feel I am to share about my recent experiences, there are some things that must remain forever a sacred trust secret between me and God. For those who don't know, I have been in the hospital almost continually since July 13th. I was recently released on August 22, 2018.

I have been home since then, however, my time has been consumed almost entirely with Doctor and related visits and doing hemodialysis in-center on a temporary basis. And for those who don't know, I had kidney failure in November 2012, and, initially I had to do hemodialysis in-center. Beginning in January 2013, I started doing Peritoneal Dialysis (PD) at home for myself every night.

Beginning in January of this year, 2018, I have been living a nightmare. Honestly, I only had 1-2 days per week the entire year that I was not in intense pain starting in January. I reported to my Peritoneal Dialysis (PD) Team, via my PD Nurse named Jackie, that I felt like I was being poisoned in January. And I reported that I thought the large cyst I had been told was on my one good, partially-functioning kidney had ruptured. In February 2018, I reported to my PD Nurse that I could taste pesticides in my mouth. And I reported that I was having intense pain and constipation. My pain ranged on a scale from 1-10 to 10 and frequently even higher as the year went on.

Every month while in the dialysis center for blood panels, I reported my concerns that my cysts had ruptured and that I was having intense pain, which escalated throughout the year to over 20 and sometimes 30 on a scale of 1-10. I reported all of this to my PD Nurse, Jackie, every month through July 2018. Dr. B, (one of my PD Team Doctors - not Dr. Braxton) did, finally, put me on the first round of antibiotics in April 2018 because I was reporting elevated fever on a daily basis that month. I found out later, that Dr. B was concerned about other numbers, but she never voiced these concerns to me.

My PD Team kept saying to each other, I found out in August, that they didn't understand why my numbers had changed so dramatically starting in January and continuing through May. When I heard this, all I could think was: "How could they not know? I was telling my PD Nurse, Jackie, every month that I thought the cyst had ruptured. I was telling them everything. Didn't they believe what I was saying? " While I had been informing my PD Nurse about my concerns that the cyst on my right kidney had ruptured since January, right through May 2018, no matter what I told her, my Doctors were never asking me about this during our monthly visits.

This lack of communication between me and my PD Team Doctors is a good example of a huge problem with medical professionals, and one of the leading reasons for malpractice lawsuits. It is a difficult problem that all medical professionals try to address, but, in my opinion, they still have a long way to go with regard to this issue of communication with patients.

Getting information from a doctor often feels like you are having to pull their teeth. They seem to regard patients as mindless automatons who don't care about their own health and trust these professionals implicitly to make all medical decisions for them, because of their high level of education and expertise.

I am not that type of person, for many reasons. I don't trust the medical profession at all. They have proven they are untrustworthy over and over again in my life, starting when I was a small child. I won't go into the details here, except to say that well-meaning health professionals have cost me many, many long years of pain, millions of dollars and countless hours of despair. God has always provided for me, but it came at great cost, great sacrifice over the years, and great time sinks, along with a modicum of despair, hopelessness and depression. But I will not be kept down for long. That is not my nature. I prefer to be my happy self.

So everyone knows, I would not have gone to my Doctor visit or for labs each month if I wasn't feeling up to it, but God helped me, and my essential oils helped me. I used massive amounts of Young Living oils, supplements and lotions to try to deal with and overcome the problems I was experiencing. Young Living Essential Oils, their supplements and many other products are a part of our daily routine at home. They keep our air clean and protect us and help us stay well. Young Living products have saved my life more than once and given me a much better quality of life. Within a year of starting with Young Living Essential Oils, I had no pain for the first time in over 20 years. I also experienced the first 12 years of perfect life I can remember in this lifetime thanks to these products, along with meditation and God. However, there are some things that require a more aggressive approach.

Somehow, with all that I was doing and God's help, I made it to my two visits each month through July 2018. I know that from my PD Team's perspective – based on those two visits, that I must have looked just fine. I was not fine. Their lack of concern, their lack of questions about the cysts was confusing to me, but I didn't mention it to them in the monthly Doctor visit because I had already told the PD nurse everything and I believed she would convey that information to the Doctors.

When the PD Doctors didn't ask about my pain levels and the potentially ruptured cyst, I thought they had all the information they required. And my mind translated the available information into this: there must be nothing to be done except monitor the situation. That was a wrong assumption, an error in logic on my part, that had catastrophic results. I should have been more pro-active about my own health care and I should have been pro-active early. Unfortunately, it never occurred to me at the time.

I felt I had been locked in my body for this year, something I had not experienced in my entire life. I could only imagine the pain others feel when they have no awareness beyond the physical before this experience. It opened my eyes to the pain of the world. Even though I meditated, and could sometimes get beyond the pain, it was so intense that I thought I would succumb. The only respite I had during this time between January and July was when I used some of the higher-meditation techniques I learned from Self-Realization Fellowship (SRF). They allowed me to get beyond the pain and feel God's Love.

I am a Christian who has been practicing Yoga exercises and meditation on God since I met my husband and he introduced me to the meditative path of Self-Realization in 1975. Christian, Muslim, Jew, agnostic or atheist, regardless of your religious affiliation or lack thereof, meditation is a way to clarity, productivity, peace and true freedom. If you would like to learn how to meditate, then I highly recommend that you learn meditation at this YouTube website: SRF-Yogananda.Org. SRF is the organization though which I learned how to prepare body and mind for meditation on God. Fortunately, you can learn the basic meditation technique on their YouTube site. Things are so much easier and much faster. Here is the link to their YouTube channel and the first of a larger series of short videos that teach you how to meditate:

https://www.youtube.com/watch?v=JytGRfZ0KVQ&list=PLCqADxk3PzviAO6qIE8KfkOJnTC16ooBy

Self-Realization Fellowship (SRF) only teaches exercises that calm the body and mind allowing meditation on God. To get the exercises and higher techniques requires applying to become a member of SRF so you can receive additional instructions via written lessons. This may not be for you, but I suggest you give it a thorough consideration if you wish your life could be happier, more peaceful and more productive. Yet, even the basic meditation technique available on YouTube will help many of you. We did not have the internet when I met my soulmate Ed and started meditating. Also on their YouTube channel, they also offer quite a number of beautiful, short, guided meditations that are very helpful.

The cyst on my kidney was finally confirmed ruptured on May 17th, by my Urologist, Dr. Greg Hallman. I had been asking his Nurse Practitioner, Michele, about the possibility that the cyst had ruptured for nearly two (2) months, every time I went in for a catheter-catch urine sample starting in April. The first catheter-catch urine test was set up by Dr. B from my PD Team immediately after starting me on the first round of antibiotics. Michelle always said she would have to talk to the Doctor and would get back to me.

Dr. Hallman's office continued with an additional 3 rounds of progressively more intense antibiotics which left me with extreme nerve pain and activated my herpes virus (a virus of the nervous system that I am told most all people carry).

The Urologist's office found that I had an extremely rare bacterium and they didn't really know how to treat it, except with progressively stronger antibiotics. This caused me great constipation and more intense pain. The Urologist's office thought they had this rare bacterium stopped with the last round of antibiotics they gave me right before my cystoscopy, but that was not the case. The bacterium re-established itself with a week of my going off the antibiotics, which occurred on May 17th.

How do I know this? Within a week of the cystoscopy, I was running a low-grade fever again and none of the essential oils that I used for infection seemed able to knock this low-grade fever. I used massive amounts of Young Living's (YL's) Thieves Essential Oil, Frankincense essential oil, and Oregano essential oil. I also took at least 4 ounces YL's NingXia Red to keep my strength, energy and stamina up. And I took massive amounts of supplements and other oils to help with everything that was happening. The list is too long to include here. Nothing seemed to work. It was truly a puzzle, but I figured it was because of the ruptured cyst. This rare bacterium had been stored in that cyst, was my deduction, stored there to protect my body, but now it was loose and causing all kinds of problems for me.

I asked Michelle, Dr. Hallman's Nurse Practitioner, if she had talked to the Doctor, several times. Her only response was to ask me again what I wanted her to talk to the Doctor about, and then saying she would have to talk to the Doctor again. How frustrating! Again, no appropriate response.

The Urologist's office left me a message telling me the cystoscopy was set up for May 17th. I called Michelle and asked her again if she had talked to Dr. Hallman about the possibility that I had a cyst that had ruptured. She said I could talk to Dr. Hallman myself, when I had the cystoscopy on May 17th. So, shortly after having anesthesia administered on May 17, 2018, I finally got to see Dr. Hallman. I explained what I had tried to explain to everyone starting in January 2018. I also explained that he had first told me I had a large cyst on my right, functioning kidney after I had an ultrasound and a cystoscopy by him in January 2017. I vented some of my frustration but tried to keep things as civil and light as I could. No sense blaming this Doctor for the failure of others. Besides, he might be able to provide me the answers I had been seeking.

Dr. Hallman listened to me attentively. He looked puzzled, then asked, "I told you that you had a large cyst on your right kidney?"

And, of course, I responded, "Yes." I also said, "You have an MRI that was ordered by my Primary Care Physician's Nurse Practitioner, don't you? Will it tell us whether the cyst is still there? I've read the report and I couldn't find anything stated about a cyst."

Dr. Hallman thumbed through the papers. He located the MRI report. He read through it quickly, then told me, "You are right. There is nothing about a cyst on your right kidney in this report." He then said the cyst had indeed ruptured and he said that most people would have died if that cyst had ruptured and it went untreated. Well, I was in intense pain for almost all of the entire year, but no one seemed to find that important, and no treatment occurred by anyone else until April. And I did almost die, several times. However, those Near-Death Experiences didn't start until

July 13th, 2018, nearly two (2) months after Dr. Hallman told me that the cyst had indeed ruptured. And yet, there was still no additional treatment.

What am I to make of this? What was the appropriate treatment? Why didn't anyone talk to me about this? I hope to be able to get those answers. However, my PD Team is still worried that I had some kind of psychiatric episode and that I'm still not in my right mind. Or that I will flip-out again, and they will have to deal with the situation.

I would like to be clear about something, because my actions may not seem rational based on what I've told you so far. The truth is, I work diligently at changing my focus from anything negative to something that makes me happy. I have always made this effort to find something to take my focus to the positive as quickly as possible. This is what you do when you are focused on spiritual growth, as I have always been. It has kept me sane and happy over the years, something I value greatly, and I expect anyone would. However, each time I faced the monthly bills it forced me to return to my frustrations and anger. And each month when I visited with my PD Team, getting no communication, I found I had even more ample opportunities for spiritual growth!

Opportunities for spiritual growth was the way I always looked at my troubles and travails. We are here, in school, to learn how to be better people. Fighting our natural tendency to focus on the negative was part of my daily approach to life. Overcoming negative impulses is what I know to be truly necessary for my walk with the Lord. If the idea that "life is a school" is a new concept to you, then I suggest you do some additional reading on that subject. I can recommend the Autobiography of a Yogi by Paramahansa Yogananda, the founder of SRF, if you required a recommended source. His easy to read book has changed the lives of millions around the world, and this concept that life is a school will change your life forever if you embrace it. For those who are concerned about non-Christian teachings, Yogananda was not anti-Christian, he worshiped Christ as the Savior. But, do not think that we are here just to enjoy everything, although that is part of what our Heavenly Father has planned for us; but first we must learn to be overcomers of negativity and fear.

There are 17 verses that I found through a quick internet search that addressed the topic of being an overcomer. This is what our Lord Jesus said:

> *"These things I have spoken to you, so that in Me you may have peace. In the world you have tribulation, but take courage; I have overcome the world."*
>
> John 16:33

We can, like Jesus the way-shower did, overcome the world. I urge you to begin practicing being an overcomer. It will change your perspective on everything that happens in life and you will begin to see God's hand in your daily life as you have never before been able to see Him.

It turns out that my first trip to the hospital was the basis of all the following physical, i.e. medical problems that followed. There was a chemical imbalance between my salt and water. This threw my electrolytes out of balance and impacted my brain. This also led to a rapid rise in my BUN, a measure of kidney function that is looked at but not as important to my Nephrologists as the Creatinine levels, according to Dr. John

Lake, my new Psychiatrist. And during this time my brain was flooded with ammonia, which led to periods of apparent delirium. We didn't know this until August 22, 2018, because there was no psychiatrist on staff at St. Francis Hospital and for some reason, they didn't have the ability to grant temporary privileges to any specialist that was not already on-staff at the hospital. What a tragedy for me.

The first time I saw the main Doctor in my PD team at the hospital when I came back to normal, Dr. Frank Braxton - who I trust implicitly, he told me it was my new PD Nurse's, Marie's, fault that I had to go to the emergency room because she did not catch the salt/water imbalance. She did not report the change in my salt/water ratio to me or to my PD Doctors. This was a total failure by my PD team. They bear a great deal of the responsibility for my multiple hospitalizations as well as the resulting hell and torture I had to endure at the hands of well-meaning medical professionals. And the last two (2) weeks were pure torture and hell for me.

Why had they done nothing about treating this ruptured cyst for so long? I don't have an answer. And this lack of treatment for a problem I had been telling my PD Team about since January 2018 was terribly wrong. My PD Team should have been doing something to help me, and they weren't. I didn't tell Ed about this because I felt it was between me and the PD team. During this 4-month period, I had multiple friends who were part of the trained medical profession telling me that I should file a lawsuit against my PD Group and go somewhere else. To my knowledge, there was no one else in my area who could take over my PD monitoring. It seemed like a "Catch 22" for me. I had to stay with them, even if their care was not what it should be. That is a problem, especially in rural America. There are often few choices.

> *"God's vastness I glimpsed in the skies of quietness. His joy I tasted in the fountains of my existence. His voice I heard in my unsleeping conscience."*
>
> **Paramahansa Yogananda**
> **Metaphysical Meditations**
>
> Page 38, 1994 Printing

It was a blessing that on my very first visit to the Emergency Room, God took me into a Near Death Experience (NDE) and a visitation from the Holy Spirit, or as the yogis who meditate on God call it - Samadhi (a Sanskrit word meaning Union with God). an extremely deep spiritual experience.

Outwardly, what happened did not look like union with The Holy Spirit or God. It looked like I was incoherent and unable to complete a task, and then resistant to my family's suggestions. My son, Joshua, had to carry me to our van and physically force me to go into the Emergency Room during the first trip. The second and third trips to the hospital were much worse.

I thought it was all some great mistake, especially the first time it happened, but everything changed when God stepped in and took me closer to Him, while I was in the Emergency Room. Ed took a lovely picture of me that he shared on Facebook with everyone, to let them know I had been admitted to the hospital. Some friends told me they were very upset by this picture because it looked like I was dead. They were so close to the truth, it will perhaps be shocking for them to know this and have it confirmed.

Here is that picture, taken at about 4 am, Friday, July 13, 2018. Please notice the serenity on my face. This was because I was already locked in my first NDE with my Beloved Lord. And, oh by the way, Friday, July 13th was mine and Ed's 43rd Wedding Anniversary. We were married on Friday, July 13th. It was an auspicious occasion for me and for Ed.

Jacqui in the Emergency Room July 13, 2018

My well-meaning family took me to the hospital to find out what was wrong, and there was something physically wrong the first time I went in – the ruptured cyst and the chemical imbalance, but the trip to the hospital and the experiences in the Emergency Room were terribly traumatic for me. It was only by God's Grace and Mercy that the flashbacks to an early childhood trauma were blocked when God allowed me to enter the first of what would ultimately be five (5) Near-Death Experiences on July 13th and my first deeply personal experience of Communion with the Holy Spirit or Samadhi Union with God in this lifetime.

There is no outer awareness during a deep Communion with God Experience. It is all internally real. [The Kingdom of Heaven is within.] However, outwardly it can be scary to those who love you and to medical professionals who do not know, but are trying desperately to find out, what has happened to you. Your actions do not match the norms. And they especially did not any of my previous norms.

I don't remember anything that occurred outwardly, in the world around me from the time I entered the deep Communion with God Experience, right after my first trip to the hospital on July 13th until the night before I returned home from the hospital that first time. Everything in between the night I entered the hospital and the night I awoke is now a total blank.

I know nothing that I said or did, except for some fixation on a pattern of behavior that God was teaching me. He said it was very important to my overall health. My only knowledge in-between entering the hospital and the first time I woke up was of the very personal experience of God. It made being Born Again seem superfluous, meaningless, and yet prior to this experience that very Christian Born-Again Experience, and I am a Christian, had been the single most important experience in my entire life. My memory returned to that of normal waking consciousness as soon as I woke up in the hospital, the very first time.

Now, God is telling me in the quiet of my mind that there is another memory I retained. It was of a dream that God showed me when He first took me into the NDE. It showed exactly what was to occur, in oblique, sepia images from old movies. It was a very strange mix of types of movies, presented as if they were a commercial or a televised teaser about what was to come.

The dream held great significance, each movie represented one aspect of what was going to occur to me over time. Each movie came with a specific message for me. And while I knew I would be coming back to the hospital, I didn't know exactly when. It could be 3 days, 3 weeks, 3 months or 3 years or more. That remained to be seen.

One of the messages given during the strange dream was that I would be attacked by negative energies or entities and they would take control of my body from time to time, making me like a Zombie. I would be talking and making movements, but I would have little or no memory of what happened at all. I assure you, I have no memory of anything that I did that appeared or was negative in any way. My family tells me there were episodes when it appeared that someone else was speaking through my mouth and acting with my body. Very scary to them! But the physical body was just acting out scenes from the "movie" God was showing me.

Another series of movie images showed me the progression of my hospital stays, and that they would be multiple. The last important message I received during the strange dream vision of old movies which came during my first NDE was showing me the way to Freedom. The time 4:44 was significant. I wrote down several times in my notes.

That one message led to an apparent fixation on hand gestures and repetitive language that was very frightening to my loved ones and hospital staff. It looked like dementia to those on the outside, in "real world" consciousness. To me it was God teaching me a new language, a code, a new way to help me integrate my personality again and communicate quickly and effectively with my family and the medical professionals. Quick and effective communication did not occur, but that was not God's fault.

This code, this new language, would help me re-integrate myself after a total of five (5) NDEs. It was clear from the beginning this was going to be a life-changing experience, and it would require many things of me that had never before been required. The Freedom at the end was all that mattered to me. That was my overall goal, 444.

The first validation that what I had experienced had been real was the first words I heard from someone in the "real--outer world" after waking up my first trip to the hospital. The words came from a male Nurse, a young man named John – who I called John 3:16 because he was not allowed to tell me his last name. Yet, he was the first man I met after my deep experience with God, and God had already told me that this man would bear an important message for me.

John 3:16 was sitting by my bedside when I came back to awareness of the outer world. I asked him what I had been saying. He responded: "Honey, being with you is like having a direct line to God." His message confirmed that I had indeed had a God-experience like none in my life before. It was exhilarating and gratifying to know that my inner experiences had been confirmed by the first words I heard in the outer world.

My Christian friends will know that most of us Christians think we get to see God when we die. Well, St. Paul said: "In Christ I die daily." This conundrum has puzzled those who do not know how to meditate. And in addition to being a Christian, I have also been practicing yoga and meditation since 1975, in order to learn how to "Be Still and Know That I Am God" as commanded in the Holy Bible. This is what God has told us we must do, but meditation is not taught in the Christian Churches that I know of, in a way that allows us to quiet the body and mind, allowing us to tune in to the Holy Spirit. For forty years, Ed and I meditated together morning and evening, deepening our connection with each other and with God.

I am happy to report that since coming home on August 22, 2018, I am off all medications, feeling much better and gaining strength and stamina. It seems a very slow process to me. However, it is progressing at a good rate according to everyone around me. So, I am happy with how things are going.

Now, why does a Psychiatrist know that this BUN parameter is so important to my health, yet he says that my Nephrologists, my kidney doctors, do not watch that as closely. According to Dr. Lake, a psychiatrist, the salt imbalance that led to the rapid rise in my BUN levels to over 90, produced periods of delirium because my brain had been flooded with ammonia. This manifested outwardly as delirium, the resulting resistance to my family and others trying to help, and a lack of ability to complete any task in a reasonable amount of time, then rejection of the need to go to the hospital. That is what my family was seeing and hearing, and what anyone still functioning in the "real-outer world" saw too. However, that is not what I was experiencing internally.

Dr. Lake told me that once the imbalance had impacted the brain, the brain required a considerable amount of time to recover. This resulted in my family seeing me exhibit the same behaviors repeatedly each time I was released from the hospital, after only a short stay at home. While the amount of time for the brain to recover varies with each individual, my nephrologists were apparently not aware of this. And they were totally confused by what happened, totally confused about why I was back in the hospital so soon, and in their infinite wisdom (sarcasm intended), they determined that there had to be a psychiatric reason for my behavior. This was practicing medicine beyond their training, totally against all medical professionalism, and yet it happened to me. And, I dare to say, it happens to others. How can I say that? Dr. Lake told us he had tried to educate doctors without psychiatric training over the years, but it wasn't easy.

This was not realized or known by my Peritoneal Dialysis (PD) Team and the only Doctor I trusted on that Team, Dr. Frank Braxton, my Nephrologist, the head of my PD Team. He told me that my labs showed everything was back to normal and in balance after my first, short visit to the hospital and I felt great when I was being released. My PD Doctors had done a great job at getting my electrolytes, my salt and water, back in balance but they did not understand that there would be longer-term effects on my brain as a result of that imbalance. So, when my labs showed everything was back in balance, they released me from the hospital and let me go back home.

The first Hospitalist that I remember seeing at St. Francis Hospital was a Dr. Finell. I saw her twice before being released from the hospital the first time, including the day I was released. I have total memory and recall of that event. It took a long time to be released from the hospital, and we weren't able to leave until early afternoon.

> *"Love is not always the perfection of moments or the sum of all the shining days- sometimes it's to drift apart, to be broken, to be disassembled by life and living but always to come back together and be each other's glue again. Love is an act of life, and we are made more by the living."*
>
> ~~ Richard Wagamese from *Embers*

Unfortunately, due to the lack of a psychiatrist like Dr. Lake being on staff at St. Francis Hospital to explain what had happened, I found myself back in the same hospital within a couple of days of my release, because my family found me standing in my office making hand gestures and repeating things to myself. I was not communicating with the outer world again, because I was locked in another experience of God. Should it be surprising that I didn't want to return to the hospital? I think not. However, it is also not surprising and not unexpected that it would happen, even by me – thanks to God's warnings. It just came much quicker than I expected.

If my family, misguided by hospital doctors' misinterpretations, had been able to understand what had happened, it would have made things much easier on us all. Unfortunately, the doctors who had seen me were not educated in brain science, and my family didn't get the necessary information about why I was still having problems until after I got out of the hospital on August 22, 2018, the day I was finally evaluated by a psychiatrist, Dr. Lake.

Outwardly my experience looked like a mental breakdown. Inwardly, the second time it happened, I was already locked in another NDE. I have no memory of anything past the time I was standing in my office talking with God. He was teaching the hand-signals and language I must use to get back to normal. He was showing me that I was going to be back in the hospital. Still time was not part of the equation. So, I was surprised when I woke up in the hospital again. This episode terrified my family, but it was all for a reason.

My second visit was fairly short. I have little memory of this hospital stay, due to my brain being damaged, however, I have great memories of God and of the multitude of people who were praying for me. Thank you so much for your prayers. You were all in my room with me. And I got to spend many hours in silence and meditation during my first two (2) hospital stays. This was indeed a blessing. During that time I was given insights into what our mission was, Ed's and mine, as a couple in a spiritual marriage. This writing and the book it will help complete, is part of that mission.

To my knowledge, the medical staff had not started doing much to address the apparent dementia, and when I came back to normal function relatively quickly, they released me from the hospital a second time, on August 17th.

My PD medical team had never told me about the change in my sodium levels prior to entering the hospital. They never even said that this could be a problem or what problem it might pose. This was a failure on the part of my PD Team, along with other failures in communication that I have already mentioned.

I had seen my PD nurse, my PD Doctors, and Dr. Hallman's Nurse Practitioner, Michelle throughout the year. I had reported my questions and concerns to all the nurses I had seen and expected that information to be relayed to the appropriate doctors. No doctor and no nurse ever communicated back to me that the doctors had received the information and what they might think about it. This is unfortunate and far too common in the medical field. Additionally, the problem with the infections resulting from a ruptured cyst on my one remaining kidney were not resolved until after I was hospitalized and received antibiotics intravenously, as well as a healing from God.

The second Hospitalist I saw at St. Francis Hospital was a young man named Dr. Mench. He told me that I had multiple infections resulting from the ruptured cyst. I had been telling my PD Nurse, Jackie, about this since January 2018. The only thing done to help this situation was done by me, myself with essential oils, until April. In April 2018, Dr. B – one doctor on my PD Team, ordered the first round of antibiotics. Dr. Mench had ordered the intravenous anti-biotics and he spent enough time with me to know that I was coherent and lucid. How much I appreciated him!

God showed me that Dr. Mench would be instrumental in my healing during my second Near-Death Experience (NDE). He was indeed! He helped me understand why I was sick, and he was also responsible later, after additional trips to the hospital, for arranging the appointment with Dr. John Lake, that finally confirmed what I knew to be true about my situation. Ed had gotten the same opinion from his research partner, Dr. Vernon Neppe, who is a psychiatrist, but he is in Seattle, and of course, doctors cannot diagnose long distance. So, even though Ed conveyed Dr. Neppe's opinion to the hospital staff doctors, it had little effect.

When Ed took me to see Dr. Lake on August 22, 2018, the day I was last released from the hospital, it was an incredibly satisfying and uplifting experience. After reviewing my file and talking with Ed and me, Dr. Lake took me off all the drugs that I had been placed on by well-meaning but under-informed doctors who were practicing medicine beyond their skill level and training. What happened to me is standard practice in a hospital when the appropriate specialist doctor is not available or does not have staff privileges at a hospital. This atrocious experience led to a lengthy and unnecessary hospital stay. I am not happy about this, as anyone might well imagine. And try as I might while I was in the hospital, there was nothing I could do to change the situation, because according to everyone else, especially the medical professionals, I could not be in my right mind, and my family had to agree.

Oh, how wrong theses well-meaning medical professionals and my family were. And how wrong my family was to believe them. But that is water under the bridge now. I know they were doing the best they could, all of them, it just wasn't the best for me, in my opinion, because it caused me so much anxiety and suffering!

I re-entered the hospital for the third time on July 23, 2018. This resulted from similar symptoms manifesting while I was at home not long after I returned from my second hospitalization. But this time was much worse. And my resistance to returning to the hospital was greatly magnified by my knowledge that it would not be a short-term stay. However, my family was worried that I had lost my grasp on reality. This was mostly because they were getting bad advice from my doctors, and they didn't really understand why I was still exhibiting such strange behaviors at home.

The hospital staff tried this time, but was not able to move me to another facility that had a Psychiatrist on staff because I was on PD for my dialysis. This went on for weeks before I agreed to have a Hemodialysis (HD) catheter placed in my body.

The few facilities who could do PD (which was required to be done by medical professionals, even though I was training or upgrading the training of almost every nurse who did manual PD exchanges for me) didn't have a bed for me. The potential facilities were always maxed out with PD patients, because so few nurses could do PD for patients. Many more options were available to HD patients.

During this third hospitalization, the doctors employed heavy medications and if I refused them, they were given to me in shots. The height of indignity! The nurses refused to listen to anything I said, and yet they agreed with me at seemingly every turn and did exactly what they wanted and whatever the doctor had ordered, regardless of what I said or what they appeared to agree to in conversations with me. I was livid!

And even more upsetting was that my family appeared to be agreeing with me, and yet also refused to do anything I asked them to do if it was not in compliance with hospital rules and policies, because they were told that if they didn't comply, Medicare would not pay for the rapidly escalating medical bills. It was incredibly frustrating.

The only bright spot in this third stay in the hospital came when my son finally agreed to let Ed bring me something to write on, so I could finally start to get some of the things that had been crowding my mind for so long down on paper. The second bright spot came when I was allowed to have some of my Young Living Essential Oils and other products in the hospital. And then, I started sharing my joy for Young Living with anyone who would listen.

The Hospitalist who was seeing me at this time was a Dr. Williams. He was a big guy with a heavy beard. After Ed and I reasoned with him, he changed my medication from Geodon (a strong anti-psychotic that was addictive) to Abilify. This medication gave me clarity and assured me that I was actually alright.

Unfortunately, I had had enough. The day after I was changed to Abilify, I told my nurses I was leaving. I was refusing further treatment and they had no right to keep me in the hospital. I packed my little backpack and walked down the hallway to the elevator. They called a code alert and surrounded with me what seemed to be 25 or 30 people at the elevator.

A big burly guy with a Security Guard uniform forced me into a wheel-chair and then there was the new Hospitalist I had seen most recently, who I really did not trust or like, standing in front of me ordering me back to my room. Someone else, who I didn't know, was standing closer to me with a big smile, telling me it was just that everyone was concerned for me and I needed to return to my room.

Another big, strong man had hold of my left arm. The first security guard had changed into scrubs and he was holding me down on the right side. I tried to reason with them. I explained my perspective. I answered their questions. They would not budge.

One of my favorite nurses, Lindsey, came and she said I had to go back to my room, but I wasn't about to do that. I knew she had been lying to me, but she didn't know I knew that. I had decided I was leaving, and they couldn't keep me here against my will. At least that is what I thought, but I was wrong. And every time I tried to move the wheelchair, the two guys holding me down increased their intense pressure and forced me to remain where I was. This action threw me into a flashback of a rape I endured by a date I had with a young man when I was only 15. He was over 21. I yelled, "RAPE!" at the top of my lungs. Nobody seemed to care. They just looked at me like I was crazy.

When I came out of the flashback, I tried to reason with them again, but they weren't listening to me. They just kept saying how concerned they were for my safety and that they couldn't allow me to leave the hospital. Lies from my perspective. They were hurting me over and over again. Over-medicating me and treating me like I was crazy, when I was not. But, it got worse: I said, in exasperation, that I would rather be dead than stay another day in this damn hospital. It was an innocuous statement of frustration. Unfortunately, it was not taken that way.

That was all they needed to hear to force me back to my room, against my will, physically put me in my bed and administer a high-level dose of something that totally knocked me out. I lost 24 hours. When I came to, I had bruises from one end of my arms to the other. Many of them had come from the forcible return to my room, but so many had already been there because nurses jabbing and seeking veins for the numerous rounds of labs ordered by different doctors throughout each day. And this continued.

Often, while in the hospital, I had labs drawn two, three or even four times in one day. I was like a human pen-cushion to these people. And I have veins that collapse and roll. They are tiny and every time they stick me, it is incredibly painful. Plus, this pain activated all the nerves that had been fried by intense pain I had been enduring for pretty much the entire year.

The hospital medical team put me on Suicide Watch after my trip down the hall, and a medical professional, a nurse, or someone was in my room pretty much 24-hours a day. This was outrageously torturous. I was a prisoner in my own bed. It was nearly intolerable. But somehow God let me know that I had to endure this torture with a modicum of acceptance. The Hospital had also installed an in-room visual monitor. If I moved or sat up in my bed, when no one was in the room, I was yelled at by someone watching me on the hospital's monitor, "DO NOT GET OUT OF THE BED, Mrs. Close. DO NOT GET OUT OF THE BED!"

I was a prisoner in my 3 ft x 6 ft bed. I couldn't get up, I couldn't go to the bathroom, I couldn't move without triggering the monitor's commanding, dictatorial response. And I hated every minute of it. I am not a passive person by nature. I do have the patience of Job according to most, but this hospital had pushed me so far beyond any red line I had that I wanted out in the worst way.

The new hospitalist had also taken me off the Abilify and put back me on a larger dose of the anti-psychotic called Geodon again. It was awful and left me with slurred speech and little control of anything I did or said. Only when the effects of this medication had worn thin did I have any semblance of real lucidity. I hated this but was not conscious most of the time I had no control. I only have memory of when I was lucid or returning to lucidity.

After several days of this, I decided to call a lawyer. I was so upset by all that was happening to me, that I was seriously considering taking action against the Hospital and even my family. How could my loving son and husband, who I trusted implicitly allow this to continue? I didn't have answers, and they were not providing any. Contacting a lawyer without discussing it with my family was not something I would have ever done in the past, but things had changed. I had changed.

My personality had been changed by multiple Near-Death Experiences (NDEs), and now I was no longer the fun-loving supporter and enabler. Instead, I became a Commander in Chief, and if you cross my red-line, I'm going to be vocal, assertive, insistent and if pushed too far I will tell you that I no longer care what you think about anything, but most especially about my health. That is my domain. "Don't tread on me!"

I contacted a lawyer and told him I wanted him to be at the hospital the following morning to help me get out of the hospital, and that I might have to file a lawsuit against the hospital and against my family. I called Ed and told him that I needed Joshua and him to be in my room at 8:00 am. This was shocking to my family when they found out what I had done. My son Joshua immediately called the lawyer after leaving me and cancelled everything, and he didn't tell me what he had done until the next day.

My son had convinced me to sign a Power of Attorney (POA) giving him say over whether I got out of the hospital, because the doctors had convinced him that I was not in my right mind. What I signed said my son would carry out my wishes as long as I was in my right mind, and it could be rescinded at any time. However, my son was not carrying out my wishes and I could not rescind the POA because of my apparent dementia or psychosis. This was not clear to me at all until that very last week in the hospital, which ended on August 17. Why my son was not following my wishes was not clear or confirmed for me until after I got out of a second imprisonment facility: the Senior Lifestyles Ward of Delta Hospital, located in Sikeston Missouri, on August 22.

It had, however, become obvious that I had to be compliant or I would never get out of the hospital. And I had asked God to protect me from myself and promised that I would accept what happened as coming from Him. So, I quelled my natural tendencies to the extent possible for me, and tried to understand things from my family's perspective. They were indeed worried about me, terribly worried, and so stressed by what they had seen and heard and dealt with night and day for the last several weeks that they could hardly stand to be at the hospital with me more than 30 minutes to an hour at a time the last week I was there.

Many hours during my first two hospital stays were spent in silence and meditation. However, I almost couldn't sleep while at the hospital at all. This lengthened the amount of time it took for my recovery from the chemical imbalance. Just for the record, I am a chronic light-sleeper and have been so since childhood when many of my early traumas occurred. I was always the first person awake at the slightest sound in my house for as long as I can remember.

The first 20 years of my life was a hell on earth. The next 40 were like being in heaven for the most part, because of my knowledge of God's participation in my daily life, and the wonderful love of my husband, Ed Close, but more about that later.

The last 5 years since I had experienced kidney failure had been a return to a sort of purgatory and more recently, since January and especially now this last week in the hospital, was absolute, pure hell for me.

I was totally lucid and not happy most of the last week I was in St. Francis Hospital. I tried to accept what was happening, but it was not easy, and I chafed at the prospect of being moved to yet another hospital. Still, I managed not to escalate to another episode, though my family thought I was still reeling up and down and looping psychotically, because the things I said still didn't seem rational to them. They were, however, totally rational from my perspective. Unfortunately, my perspective held no weight at this point in time. It was absolutely the most horrible nightmare I had ever experienced.

Dr. Williams, one of the Hospitalists I liked, had been in a slightly contentious, yet friendly relationship with me for several visits. He had reduced my anti-psychotic medication and the 24-hour time loss and mental looping had stopped. Even the reduced medication made me terribly sleepy, but as far as I could tell, other than when drifting in and out of consciousness, I was totally lucid again. Unfortunately, no one accepted this as true, and nothing I said made any difference.

It was so frustrating! If someone else forgot something, it was normal, but if I forgot anything it was a big cause for concern. If I said something they didn't understand, it was because I wasn't in my right mind. And the nurses and lower-ranking staff had been told to agree with anything I said just to keep me calm. I trusted many of them because of this but found out either before or after I was released from the second facility I was sent to, on August 21st, that these lovely nurses and other low-ranking staff had been lying to my face, for the most part. How could they do this? They were following orders. They were trying to do what they had been told would help me.

I begged my loving husband Ed to get me out of the hospital, and he wanted to, but he couldn't. We didn't have the money to pay the mounting medical bills, and Medicare wouldn't pay anything if I refused medication and left the hospital. He said the hospital would not release me to him without a psychiatric examination by a qualified psychiatrist, but they had no psychiatrists on staff, and none were available locally, and every facility with a psychiatrist within a hundred miles, either had no beds open, or no PD dialysis capability. Catch 22!

The hospital had tried for two weeks to move me to another facility with PD capabilities and a Psychiatrist. Nothing had opened up. Dr. Braxton told me the week before my final exit from St. Francis Hospital that if I would consent to getting an HD catheter placed on a temporary basis, then they might be able to move me to a facility with a Psychiatrist sooner. So, I agreed. I told Dr. Braxton about the surgeon who had placed my PD Catheter. Dr. John Foley had placed my PD catheter in January 2013. Dr. Braxton had Dr. Foley come see me. He would place the new HD catheter and, after we talked, he agreed to leave my fully functional PD catheter in place, even though it was not the usual procedure. I had been told that if my PD catheter were removed, I could not return to PD, or at least, most certainly would not be able to return to PD for several months. That would mean every-other-day trips to the hemodialysis center; making normal life impossible.

Unfortunately, when I went down to the operating room for surgery, Dr. Foley wasn't there. Another doctor from Foley's practice was, and that doctor was determined to remove my PD catheter. I argued with him vociferously, and finally convinced him before they put me under, not to remove my PD catheter, or to at least to wait until Dr. Foley arrived. This surgeon was a member of Dr. Foley's group, but he was an ego-driven jerk who treated me with disrespect and basically did whatever he wanted to me, even creating a great deal of pain for me when he inserted an IV that was a large-gauge needle into the vein in my left hand. I wish I could have punched him in the face, but that wouldn't have done anyone any good, least of all me.

Dr. Foley was on an emergency call at another hospital, I learned as I was going out, but he arrived just in time. And I was saved from the irreversible actions of this misinformed (I prefer the term Stupid) surgeon who didn't know me or what was really good for me personally, at all. Medical professionals are always stuck on "this is the way we always do it", and there are no exceptions. In my opinion, there should always be an openness for exceptions, and the patient should always be informed and listened to. Otherwise, there is no consideration or personal treatment for the patient.

In my experience, it always took me about a year of working with a doctor to establish a good understanding relationship. Dr. Braxton and I had this kind of relationship. I knew I could trust him, he knew he could trust me. He also knew I was not the average, ordinary patient. He also respected Ed and listened to his questions and suggestions. This was not the case with any of the other PD Doctors on the PD team who are all in Dr. Braxton's group, or any of the doctors I saw during my stays in the hospital, except Dr. Foley. We had a good relationship, and even though we hadn't seen much of each other, Dr.

Foley treated me with respect and dignity. I appreciated him immensely as a doctor and a human being

I didn't have much trust with the other doctors on my PD team, and the doctors at the hospital, and they didn't trust my judgement on anything. This was a huge problem, and the main reason for my PD team's failure and the hospital doctor's failures too. None of them trusted my judgement even before I was exhibiting dementia or delirium or was put on suicide watch at the hospital. They didn't take what I said seriously. And they didn't like that I wanted to know all of the options and argued with them. But that was me, and will always be me when it comes to health professionals. I am not a mindless automaton who trusts them to make appropriate decisions for me. I am reasonable and open to discussion about a subject, but don't tell me you know best, because I know my body a hell-of-a-lot better than anyone else ever can. So, get used to it! That was my attitude.

Ed told me that I was never over the top talking with doctors or nurses, while he was there; I never spoke in anger, and no one would have known how exasperated or angry I was about the treatment I was receiving at the hands of well-meaning medical professionals who were treating me based on the norms of their experience with hundreds of other patients. I was good at hiding my feelings, from years of practice since childhood, but that's another story that will have to wait for now.

I was not the norm. I was not a statistic, and I wasn't going to allow any doctor to treat me that way. I was assertive and argumentative, but I think respectfully so. I knew they were highly educated and skilled at their profession. But they were not me. It was my body, and I was my best advocate for my health and that meant I had to argue with my PD Team Doctors and other doctors on a regular basis. I wasn't afraid of doing this, but it caused problems for me until we developed an understanding. Usually, as I said earlier, that understanding took about a year of working together on my health. That amount of time was not possible in the hospital. It was incredibly frustrating. And I rarely saw the same doctor twice, because hospital policies and procedures prevented that from occurring. Dr. Williams was the one exception in St. Francis Hospital. I saw him five or six times in a fairly short period of time. This was helpful, but not perfect. Thankfully, he made changes to my medication that allowed me to be more comfortable and lucid.

Believe it or not, lucidity generally ended my NDEs. I did retain a global awareness, an expanded awareness of things around me for a while, even when lucid. God showed me deep, hidden codes in everything, even in the most inane kinds of things, such as commercials, the news, and regular TV programs. It was awe inspiring to know about these deep codes. However, I found that talking about them with others led to their being certain I was still not in my right mind.

God had told me to be boldly, blatantly, brutally honest with everyone. It was painful for me at first, but I did as God had directed. This led to a lengthy stay at St. Francis Hospital, and to my son's distrust in my judgement even when I was totally lucid again. How frustrating! Every step looked like a step in the wrong direction. I couldn't keep going this way. Then the word came that they had finally found a facility in Sikeston to transfer me to that had a psychiatrist. Finally, there would be a change.

On the day of my release from St. Francis, Dr. Williams saw me and Dr. Ben Lamphor, a PhD in Psychology. They both said

I could be released to go home. Unfortunately, Dr. Makapati from my PD Team and even my own beloved son, Joshua, did not agree with this. He was afraid I would become incoherent at home again. They required that I be transferred to the new facility. They were afraid I would be psychotic and wind up in the ER again, in a vicious cycle.

It was a long trip to Sikeston from Cape, made longer because my son refused to listen to me or Ed about a short-cut to the facility that we knew. He didn't trust our judgement. Instead, he was relying on Google Maps. And how stupid! We went two-thirds of the way around Sikeston on the Interstate 55 and then 60, and then back-tracked north to the hospital which was in the north end of town. I expected to see a Psychiatrist as soon as I got to the new facility. That did not happen. In fact, I didn't get to see the Psychiatrist until late on my second day there.

As soon as I had talked with the Psychiatrist at Senior Lifestyles (a euphemistic name for an insane asylum), he took me off the heaviest anti-psychotic, Geodon. However, he ordered that I continue taking two milder anti-psychotics, Abilify and Atavan, on a daily basis. I had to comply with the administration of these medications, but I was happy to find that, fortunately, they didn't put me into memory loss, time loss, or repetitive loops. So, they were much better than what St. Francis doctors had forced me to take. If I refused an oral medication at St. Francis Hospital, they gave it to me anyway as an injection. This over-medication led to most of the confusion that kept me in the hospital for so long, at least that was my perspective.

It wasn't until I was released from Senior Lifestyles, and had seen Psychiatrist, Dr. John Lake, and had actually returned home, that I learned from my son, Joshua, exactly why he had required that I go to Sikeston. He wanted someone to tell him what exactly could be done to prevent me from coming out and then having to go back into the hospital again. He didn't want to have to go through that nightmare again, but no one could tell him that until I saw Dr. Lake. Irony of ironies, I had to be released to get this information, and yet they were keeping me in the facility at Sikeston because they couldn't get this information. No one could explain to them why I had returned to the hospital twice before. How frustrating for them.

The whole ordeal had been a tremendous strain on Joshua and on my beloved husband, Ed. Ed almost had a nervous breakdown, watching what I was going through. I can imagine how devastating this experience was to my sweetheart. I know how hard watching him go through something like this would have been for me. Although I knew this was incredibly hard on both of them, still I couldn't understand why they seemed to think I should be perfectly fine with staying in the hospital with no understanding of why they weren't allowing me to go home. Why didn't they trust me? I had no answers and that was the most frustrating part of this whole experience. I had to accept it and find a way to deal with it. It was not easy and certainly not my preferred way to do things. Ed told me later that he did tell me why he couldn't just bring me home, but I do not remember that. It apparently fell into one of the memory gaps I had for some reason. I know now that the frustrations of this experience were spiritually necessary for my ultimate good. But I did not know that at the time.

My family, like the medical professionals, no longer had a good handle on how I was doing at any given moment during those last two weeks, and they didn't trust my judgement. Instead, they listened to the doctors and seemed to me to be more interested in supporting them than me. This was incredibly frustrating. The doctors saw me less than either Ed or my son, Joshua, did. How could my loving family listen to them and trust them, but not me? They seemed to trust them to know what was best. I knew better. I had endured torture and pure hell for nearly two weeks at the hands of well-meaning medical professionals. They were constantly doing unnecessary things throughout my stay in the hospital in Cape Girardeau, and fortunately this occurred to a much lesser extent in the Sikeston facility.

At St. Francis, they insisted that I take medications I did not need, and often injected those medications into me without my permission, even after I had refused them. This was pure hell and put me into PTSD responses based on early childhood traumas, caused lost time, and allowed negative energies to take over my body for short or long periods of time. It was especially hellish the last week I was at St. Francis Hospital. I was totally lucid during that last week and every day was like being in prison. They imposed a 1-liter water intake restriction, and that, along with the fact that I was a prisoner in my own bed, I hated every minute of it. The water restriction was poorly administered, sometimes I got less than 1 liter of fluid per day. Nurses forgot to bring my ice or water, and the next shift thought it had already been done. Again, it was a living hell. But I am a strong person. I am an overcomer. And I knew I had to be compliant or I would never get out of the hospital. So, I endured.

This is what Christ asks of us, to endure that which we cannot change. St. Francis's Serenity Prayer specifically spells that out. The irony was not lost on me, but it seemed totally obscure to others who visited or talked with me during that time – except for my friend Vangie Jones. Vangie was God's lifeline for me.

Taking all the prescribed medications was totally against all my principles in life. Ed and I always avoided prescription medications like the plague because their multitude of side-effects are often much worse than the original problem. And what do they do to counter the side effects? Prescribe more drugs, and then more drugs to counter side effects of the secondary medications. It's a vicious cycle, and one that we see repeated over and over again with the elderly. Such a horrible tragedy! But this is all these "stupid" western-trained allopathic doctors know. They are not schooled in alternatives and options, for the most part. That is beginning to change, but it is still far too lacking in modern medicine.

At the Sikeston facility, I had more freedom of choice and less restrictions. However, I had to endure wanderers who couldn't remember their own names or room numbers. They would come into my room, get in my bed, or my roommate's bed, if I couldn't dissuade them. They would change our bed clothes, thinking it was their job. They would be argumentative and angry when I tried to get out of my room. These people were truly out of touch. I was not. And I helped the nurses corral these other patients and talked with the patients in comforting tones and got them to change their behavior and they began to follow and trust me more than anyone else there. They followed me everywhere. They wanted to be with me all the time. It made them feel safe.

Three of these totally out of touch people got to go home before I did! What a frustration! Why did that happen? Maybe they had already been there for a long time. I wasn't sure, but still I had to comply and be complacent to the greatest extent possible, or it would only mean a longer stay at this insane-asylum called Senior Lifestyles.

I had one goal and that was to see Dr. Lake on August 22nd, and I had to be out of the Sikeston facility by then, but the second Psychiatrist I saw at Senior Lifestyles, who took over on Monday after I was admitted on Friday, August 17th, wanted to keep me at least 2 more days, possibly longer, for observation. Somehow, I had to get that changed. I had to be out of this facility on or before August 22, so I didn't miss the appointment with Dr. Lake. That was my new red line.

Fortunately, I didn't have to take and wasn't forced to take all the numbing medications prescribed by hospital doctors, at the Sikeston Facility. So, I began cutting things out that I knew I didn't need right away. The new medications were not so overwhelming and that was good. I also found the looser restrictions very helpful in maintaining my personal equilibrium. And I was allowed visitors on a daily basis for the first time in nearly a month.

Vangie Jones and Dena Klingel were incredibly instrumental to my healing. They were shown to me in all of my NDEs as two of the numerous people who would be God's emissaries and would play a very important role in my healing. That was all true.

I had begun to doubt my own sanity after I entered Senior Lifestyles facility. It was depressing. I began to think maybe the only reason I had any lucidity was because of the medication. This was not true, but it caused me great emotional pain because even I questioned myself now. Fortunately, Vangie and Dena lived nearby.

The visits with Dena and Vangie changed everything. They reassured me and helped me know, just by spending time with me, that I was okay. They listened to me intently. We shared much with each other about using essential oils and other alternatives to help with the problems like those I was having. Dena employed massage and oils to help alleviate some of the problems and pain caused by the overload of fluid that I had to endure between HD treatments. Vangie and I shared so much on such a deep level. How I appreciated the time I had with these two wise women. They saved my life, they saved my sanity. They knew that I was not crazy. This was so important to me.

Vangie also brought me one of the most important and exceptional books I have received in years. It's called: **The Circle Maker, Praying Circles Around Your Biggest Dream and Greatest Fears**, by Mark Batterson. If you haven't ever read this book, and especially if you are a Christian, then look for it and start reading it right away. Vangie put my fears and concerns in her prayer circle. She prayed for me daily. She helped me in ways I cannot begin to explain. But just know that I have the deepest love and admiration for this wonderful woman, who is a wholistic nurse with great understanding and experience in life circumstances.

Vangie retired from being a hospice Nurse after 20 years woring in that field, and then became a holistic nurse and is sharing the life-affirming message of Young Living with anyone who will listen. She is an awesome leader in our Young Living (YL) organization, and I am proud to call her one of my dearest friends. While we have had so little personal contact until the last few weeks, she has been incredibly important in my healing process. I LOVE GOD and He never fails me, and he showed me that Vangie and Dena were trusted emissaries sent by Him. Oh, what JOY!

By the way, everything that God showed me in my NDEs has all proven true so far. God Alone was with me throughout, but so were many of you who were praying for me, and also Gary Young – the founder of Young Living (YL) - who passed away earlier this year. Gary was with me throughout much of the inner experiences with God and Gary was instrumental in my healing and my path to Freedom. (I hope to share more about this in future posts and publications.)

I am happy to report that I have been off **all** medications since seeing Dr. Lake on August 22nd, except for taking a generic form of Atavan which Dr. Lake suggested might help me sleep when I first got home. He told me, if this medication wasn't needed, I didn't have to take it. So, I only took it twice. Then my Melatonin arrived, and I switched to that for 3 days. I decided not to stay on the Melatonin unless it was required, because taking melatonin can affect our body's facility to produce melatonin naturally. I did not want to reduce my body's effectiveness.

I am also happy to report that I can rest assured that the diagnosis I received from Dr. Lake is correct from a physical and psychological standpoint. I had a chemical imbalance that led to damage to my brain which took a while to heal. Everything beyond that was unnecessary medical intervention. That is very good to know, and while my stay in the hospitals was overly and unnecessarily long, I know it also served God's purposes in my life. I don't understand all the purposes it served, and I certainly was not happy about being in the hospital so long, but there were good reasons for it, both inward and outward.

The previous medical diagnosis, as mentioned earlier, was delirium, but it was not a normal delirium. It was a Holy Ghost experience. I prefer to call it a God-experience. And God protected me from the trauma that my body must suffer due to the chemical imbalance and also at the hands of well-meaning medical professionals and family, who only had my best interests at heart. However, their interference with my experience of God was the most frustrating of all. God gave me multiple Near-Death Experiences that allowed me to experience total and complete Union with Him in this Life Time. It is a blessing that few ever know.

I wish I had a recording of what I said to John 3:16, aka John De Marie, but I have no memory of saying anything outwardly. I was engrossed in the Holy Spirit and that very personal memory cannot yet be shared. It is such an incredibly special gift to me to have had this deep experience of God and what I received is far too personal to share in any form at this time. It is a sacred trust given by the Holy Spirit to each person that they must save some secrets that are only to be known by the person and God Alone.

As you might well imagine, trying to explain what has happened and how I experienced all of this will take time. I am expecting, based on my recent experiences, that most of you will respond to me in basically the same manner you always have. And that's good. I wouldn't want it any other way. However, no one will think I am the exact same person I was, even if they haven't read this material.

Most of my closest friends, and especially Ed, my beloved husband, understand or are trying to understand what has happened to me, based on our conversations and what I have made available to them in written form. Others will understand as time goes on, and more information is made available. Some will never understand, at least not in this lifetime. And that's okay.

I'm not here to change what anyone else believes. However, I have to do as God has directed me, I must share my truth boldly, with honesty. This is not easy for me to do. I have always tried to spare people from anything that might cause them to question their own beliefs or mine. I preferred to hide in partial truths so that friendships could develop naturally, rather than being changed quickly when too much information is given too fast. I had learned to do this since early childhood. As I share my stories further, you will understand how that came to be. For now, just know that I never meant to deceive anyone, but only to make you more comfortable and to allow us to get to know each other without ideas or belief-systems creating a barrier between us.

Being boldly, blatantly and brutally honest was not my role in the past, but my assignment has changed. What I post will be challenging for many, especially many of my Christian friends. I know this because of responses I have already received in conversations with one of my closest Christian Friends. It concerns me deeply that some of my closest personal friendships might have negative reactions to the information I will be sharing. I know that talking about reincarnation creates problems for some Christians, because this teaching was removed from Christian teachings during the time of the Roman Emperor Justinian 1st.

Reincarnation was once part of Christian teachings, and my beloved husband, Ed, can speak to the history of that very well. You can visit his blog for an in-depth discussion of the history of Reincarnation and the Christian Church. Ed's blog post is extremely well-written and is a quick read, even though it is long.

The information will ultimately be available in a book that Ed and I will write together that will be available sometime in the near future. That book will relate more of our understanding in a longer form. However, I would encourage you to read the post on Ed's blog. It is highly informative and extremely well-researched. His blog is called Transcendental Physics and the address is: www.ERCloseTPhysics.com

NOTE: My hope is that you will read this description of my ordeal in its entirety and also do the same for each subsequent publication of this information. If you are not a scientist, but can stay in what most of my newer scientist friends call the logical part of the brain, then we will get to a point of understanding much more quickly. The writings of the following scientists will help you if you require a greater understanding of Near-Death Experiences (NDEs) and some of what has happened to me. If you know or have heard of Dr. Ed Close, Dr. Vernon Neppe, Dr. Gary Schwartz, or Dr. Ian Stevenson, then Great!! If you need a deeper understanding of what occurred to me than is provided here, please visit some of their posts on YouTube for a quick overview of these men's works.

I expect that some of you, who have been involved with me throughout this experience, will already have some or perhaps total understanding about my situation. In fact, one very close Spiritual Sister in Washington, DC, Beverly Bachemin, experienced a great deal of my Holy Spirit Experiences with me, because she was praying for me intensely and we have been on the same path for our entire lives, though we did not meet until I was in my early 20s. Ed and Beverly had been friends for years, as members of SRF, when he introduced her to me in 1978. I have had my closest spiritual relationships with my family and those who went through this with me and two of the scientists mentioned above during the past 10 years.

My Spiritual Sister Beverly from Washington, D.C. helped me more than she could possibly know. I had called her to thank her for sending a beautiful flower arrangement to me while I was still in the hospital. When we started talking, she was telling me about being with me during my God-Experiences. She had seen the saints, sages and multitude of people who were praying for me. She confirmed everything I knew to be true about my experiences. What a joy to know that God so loved me that He would allow her to experience this and retain the memory and retell it to me in great detail. She named the saints, sages and highly advanced souls that were there praying for me, along with many of the other people in my room at those glorious times of union with God. What Joy! I am eternally grateful to my beloved God and to this wonderful Sister in Christ and a Yogini who was sent by God to empower me with the truth about my experiences.

My great thanks to those who were instrumental in my healing since going into the hospital. God showed me they would be there as His emissaries when I needed them. The few who were allowed to visit me in the hospital were so helpful and your time was so much appreciated. Thank you again.

Now, God has given me this new assignment. He has made it clear that I am writing this information about my experiences for public consumption. I have to be able to share my story publicly, because it is part of the mission Ed and I are here to complete. God tells me I must share as much of it as possible, right away. The memories of my experiences with God have faded since I returned home and have been consumed with outer things, especially daily dialysis and doctors' appointments. There has been little time since I returned for meditation. It has taken a toll on me, but I know that all things are in God's hands. Nothing is ever lost, even though it may look that way temporarily.

I doubt that anyone else goes through this type of experience the same way I have, because God makes all things unique and appropriate to each one of us. And while I have had 4 previous Near-Death Experiences (NDEs) without ever understanding why, this time, in a most unique way and really after a total of five NDEs in one month, I have understood why this happened in exactly the way it did for me. It was all a blessing from God. And, truly, the time with God afforded clarity and understanding about my role in this life with my love of many lives, Ed, as well as this particular situation, that could not be known otherwise. It is overwhelming to realize what God has done for me. I wish I could tell you everything now, but that simply is not as easy as it sounds.

I must also thank our friend Lee Lawrence for me helping me know what goal I was to focus on during my NDEs and recent visits to the hospital. Lee gave me one word before I went into the hospital on July 13th. That word was: "Understanding." How important this was for me. It was a guiding light for me.

My soulmate Ed led me to my greatest spiritual teacher in this life, Paramahansa Yogananda, who said: "In all your getting, get Understanding." And HOLY COW GOD! this experience of five NDEs in one month happened so quickly! And it was so intensely personal that I struggle to share the details with others.

I had been totally frustrated by the fact that my son Joshua didn't want to hear this information when I first became awake and had total clarity in the hospital the very first time. And why I can't remember so many things now is also concerning to me, because from my perspective, the memories of the dark behaviors or strange behaviors, the times when I was not in residence in my body, are gone in total -- disappeared as if they never existed. For that, I am immensely grateful. Details of what I did during those times of incoherence come out slowly from my family. They are embarrassing to learn.

I know that what was lost to my waking consciousness is, for the most part, a blessing. There were many traumas in my early childhood and teen years, and there were also the negative entities and energies that were inhabiting my body when medical staff drugged me to help sedate me. These memories have been buried deep by God to avoid hurting me. My family has told me some of the things I said and did during some of my delirium states. They are very ugly. This was the result of my body being controlled by negative energies, and also due to the flashbacks and experiences of childhood traumas.

My sincere apologies to anyone who was hurt, frightened, or terrified by something I said or did during these states. Pharmaceutical medication opens us to negative energies, especially over-medication, and I was being over-medicated by well-meaning doctors on a regular basis. It was sometimes very ugly. I can only tell you that I have absolutely no memory whatsoever of saying many of the things that my family has told me I said. Thankfully, I only have memories of God, the Holy

Spirit in communication with me, and the prayers and amazing blessings of so many saints and ordinary people. You filled my room every night. It was a magnificent and awe-inspiring experience. Thank you. Your prayers contributed to my great spiritual reveries, and eventual recovery from life-threatening illness. Thank you a thousand, thousand times. You contributed to my healing in ways you cannot begin to imagine. I can never thank you enough.

Dr. Lake, the psychiatrist who saved me from the hospitalist medical community, said the loss of memory was the brain's way of protecting me. However, I know that it was God was protecting me and, if I need to know something, He will make it clear and available to me again. I have more memory of my childhood than ever before, at least since coming on the Spiritual Path in 1975. This was when I was Born Again, after days of fasting and praying to my Beloved Lord Jesus Christ, who led me to Ed Close, and through him to the great Yogi and Avatar, Paramahansa Yogananda.

God blessed me as Jesus, by taking many of the traumas of my childhood and burying them so deep that they have no impact on my daily life. It is as if they were washed away by Jesus, my beloved Christ, all in an instant. They no longer exist for me. Praise God!! It is so freeing!

And Freedom is my number one requirement for this life. My beloved Lord is always watching over me, and I praise His Holy Name. Tears flow unrestrained, as I remember how He and others have cared for me through so many trials and tribulations. My Lord was always there, holding my hand and carrying me through – it is so overwhelming to me to even think about. My Beloved Savior Jesus the Christ and my great Gurus of the Self-Realization path could not prevent what the law required be fulfilled, but still they carried me through with Grace and Mercy. Otherwise, I would not be alive today. And it took them all, along with all of you who prayed for me. Thank you again. Thank you over and over again.

I mentioned earlier that in addition to being Christian, I am a Yogi or, as a female, I would technically be called a Yogini. I have been practicing Yoga since 1975, when I first met Ed Close in this life. Because of this fortunate re-connection with Ed, my loving companion of many lives, I also received initiation into the Self-Realization Fellowship's technique called Kriya Yoga in October 1977. So, I have been practicing the presence of God and preparing for this beautiful experience of the Holy Spirit since that time – over 40 years.

I also spent my youth and early adult years totally engrossed in my Christian Faith, beginning when I was only two years old. My maternal grandmother was the choir director and organ player for her little country church. She was a born-again, Southern Baptist. She put me on the front row of her choir when I was only 2 years old. I was never afraid to be on stage from that point on, because I had so many wonderful people praising me for standing up in front of the whole church at such a young age. And I remember being so proud standing there at the front of the church. I so loved to sing to my Lord Jesus. It was a great church, filled with the Holy Spirit. What a blessing to have had that experience so young.

Now, unless you have been on your own Spiritual path, and are in a close personal relationship with me, perhaps on the same spiritual path with me, or in a familial relationship with me, then what I share will here may seem quite strange. Your response will always be colored by your own belief systems and resulting interpretations of the information given. Again, I am not here to change what you believe, but only to share with bold and brutal honesty the truth as I know it. What you do with the information is totally at your discretion, your choice is all important in this sharing.

If you are in close spiritual proximity and on a similar spiritual path as me, then what I have learned may seem obvious or totally unnecessary to you. For others, what I share here will seem quite strange. Some of you will understand immediately, others will never understand any of this. That's Okay!

It is true that I have changed, even changed my personality type from being a Fun-Loving Supporter to a Commander in Chief. My only clue to why I am so different is that God blessed me. I am Blessed and Highly Favored, as are you all.

You may find reading this information frustrating, or it may put you in overload. As one close personal friend said to me on the phone, what I'm going to report may well be more than any single individual would like to see, hear, or know. I am not responsible for your reaction to what I share. My only responsibility is to my Beloved God and His command that I share with humbleness, boldness, blatantly, and in total, brutal honesty exactly what happened and what I have understood from it. I hope it serves His purposes in your life, as it has in mine.

I am choosing and desiring more than anything to maintain my beautiful and loving friendships with those of you who are close to me. I would prefer not to do or say anything that might damage our relationship. However, I must adhere to God's command to share with brutal honesty the truth as I know it.

It is true for me, but it is not required that it be true for you. Nothing is required of you, though I would admonish you to "Love God with all your heart and with all your soul and with all your mind and all your strength." Mark 12:29 This is the greatest commandment ever given by God through our Lord and Savior, Jesus Christ.

Now…If you can find no love for God or Jesus in your heart, then just focus on Love for those of your fellow human beings who have earned your trust and love. The love of all humanity, or even part of it, was also included in the scripture Mark 12:29

The full scriptural reading is Mark 12:28-34. I would ask that, regardless of your own religious beliefs, that you read these scriptures. It is one of the most beautiful scriptures in the Holy Bible, one of my favorites. It will help you understand me, because it is what I have strived for my entire life. I am not yet perfect in my practice of the truth taught in the scriptures, but God does not require perfection from us, He only asks that we do our best.

Mark 12:28-34 King James Version (KJV)

[28] *And one of the scribes came, and having heard them reasoning together, and perceiving that he had answered them well, asked him, Which is the first commandment of all?*

[29] *And Jesus answered him, The first of all the commandments is, Hear, O Israel; The Lord our God is one Lord:*

[30] *And thou shalt love the Lord thy God with all thy heart, and with all thy soul, and with all thy mind, and with all thy strength: this is the first commandment.*

³¹ And the second is like, namely this, Thou shalt love thy neighbour as thyself. There is none other commandment greater than these.

³² And the scribe said unto him, Well, Master, thou hast said the truth: for there is one God; and there is none other but he:

³³ And to love him with all the heart, and with all the understanding, and with all the soul, and with all the strength, and to love his neighbour as himself, is more than all whole burnt offerings and sacrifices.

³⁴ And when Jesus saw that he answered discreetly, he said unto him, Thou art not far from the kingdom of God. And no man after that durst ask him any question.

So, that's it for now. I am laying the basis for what is to come. Till next time, this is the new and improved Jacqui Close signing off. God-Willing, I will be able to continue these communications until the main story is complete.

God Bless You and Your Families,

God Bless Your Life and Those You Love.

This I pray in Jesus Holy Name,

AMEN!

Jacqui Close

August 28, 2018

Commentary:

Looking back over Jacqui's life of devotion and joy, so often interspersed with intense pain and suffering, I am reminded of Saint Teresa of Avila who said: *"God, if this is how you treat your friends, it's no wonder you have so few!"*

After Jacqui was finally discharged, from the hospital, she worked at decoding her notes and writing about her experiences in the hospitals, but she became increasingly involved in preparing for her departure from the physical body, and wasn't able to elaborate on her experiences much. I've searched her computer files for additional notes and writings about her experiences while hospitalized, and the one presented above seems to be the last and most complete one. The notebook she used while in the hospital, a notebook from the *"Live Your Passion" Young-Living 2016 International Grand Convention*, the last YL convention we attended, contains a few coherent comments relating to her experiences while in the hospital, with dates and times, but there are more than 50 pages of the notebook literally covered with hand-written notations in code. Those pages of code also include dates and times, and many groupings of three letters or numbers, like DNR (Do Not Resuscitate) TDP, TDL, JCP, TMC, !!!, ***, Ssc, Ss2, etc - and one three-character grouping, sometimes looking like YYY, sometimes looking like 444. Is found repeated a number of times. This may be significant, as one of the predictions she says God gave her, because her final exit from her body occurred at exactly 4:44 pm, December 15, 2018.

From this most recent file I found in her computer, it is clear that she intended to share more of what was revealed to her, but during the months of September, October and November, her priorities changed markedly. She began preparing us for the final act. She was working on things related to our finances, insurance, the Young-Living business, with our accountant, and she contacted a lawyer to prepare a will. She was busy with a lot of seemingly small details, like telling me what she wanted to be dressed in for her funeral visitation, selecting special Christmas gifts for special friends, and posting beautiful thoughts, pictures and videos on Facebook. It was like the final moves of solving the cube: She was getting the last pieces in place, so that everything was aligned. Toward the end, we began our morning meditations together again.

All the time, she was in terrible pain. She was losing weight, sleeping no more than an hour at a time, day and night. Her red-blood-cell count was low, and she had to return to the hospital more and more frequently for blood transfusions. She would perk up after each transfusion, and work on making sure we would know what to do and where things were, when she was gone. Finally, on Monday, December 10th, we met with Dr. Braxton. He outlined our options for keeping her alive. One was to remove her bladder and colon to stop the blood loss from bladder and colon cystitis. Another was selective radiation. She rejected both. Dr. Braxton looked at me with sad eyes, almost in tears, turned back to Jacqui, and said: "You'll be going back in for blood transfers more and more often. You'll get weaker and weaker; dialysis can't keep you alive forever."

Calmly and matter-of-factly, Jacqui said: "*I choose to discontinue dialysis and go on hospice.*"

I knew from mid-October on, that she had been asking God every day: "Can I return to that wonderful place in your presence? Can I come Home now?" We would meditate together each morning after she got off the machine, and she would tell me: "God says no, I have more that I have to do." Finally, on Friday, December 14, after only three days on Hospice, she told us:

"He says I can go home tonight!"

But that night went by slowly, like so many others had gone before. In the morning of the 15th, in unbearable pain, she decided to take a morphine pill. She raised up on an elbow, and I placed the pill in her mouth and held a bottle of water steady, so she could swallow. She relaxed, and a sweet smile spread across her face. I felt a warm glow as she said: "I love you so very, very much! I want *your face* to be the last thing I see, before I see the face of God!" And she closed her eyes and went back to sleep.

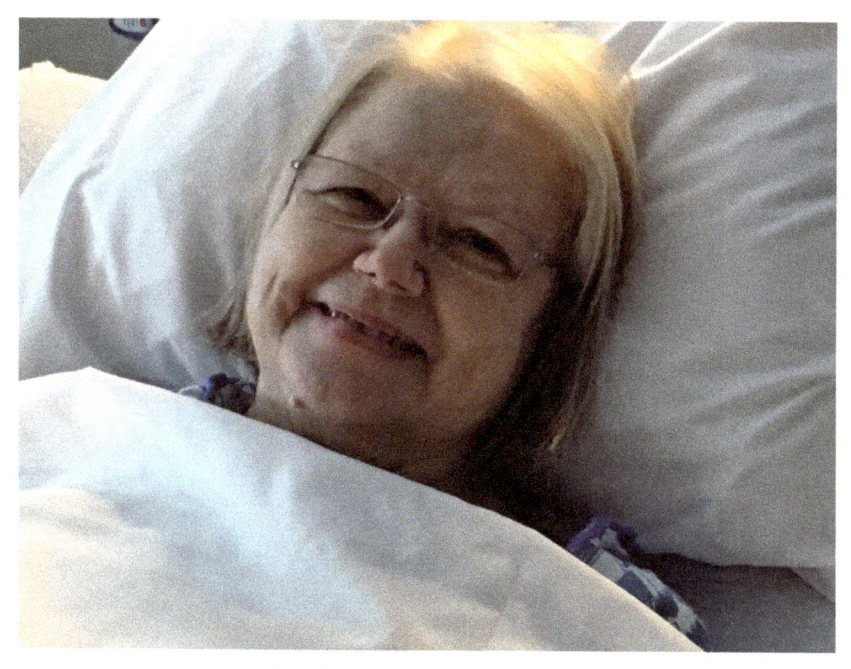

Jacqui after her Near-Death Experiences

She slept for six hours. Knowing that God had said she could go, I sat by her bedside, keeping vigil. At around 4:40 pm, she woke, opened her eyes and said: "Why am I still here?" She wanted to get up, to sit in her reclining chair by the side of the bed. I called Josh, but because she had no muscular control of her legs, we couldn't get her into a standing position, to turn and sit in the chair. I said: "Just sit here, on the side of the bed." But she couldn't even sit up by herself. I put my arms around her and suggested that she lean against me. She turned and looked at me, and our foreheads touched. She closed her eyes and I began chanting the name of *Babaji*. She breathed a few more times, and Babaji took her home. As she had wished, my face had been the last thing she saw of this earth.

THE COSMIC LOVE STORY
(A Poem Dedicated to Jacqui)

The blackest black of deepest space
Is gone when sparks one flash of light.

The blue and green of Earth do grace
The fleeting dreams of inner light.

Call up Akashic memories
Of all the histories of Earth,

The dance of all Realities
Both of sadness and of mirth.

The Light that glows behind your eyes,
The Love that burns within your heart,

Did they not know the ancient skies?
Did they not always play their part?

Imagined differences we see
Of me and you, of you and me,

Like rivers flowing to the Sea,
They merge as One eternally!

Cosmic Love knows no bounds.
As we make our Cosmic rounds:

Whether we meet again as King and Queen,
Or as humble paupers with lives so mean,

As Cleopatra and Anthony,
Or as Tristan and Isolde,

As Radha and Krishna, as Sita and Rama,
As Rumi and Shams, or Ferdinand and Isabella,

As Romeo and Juliet, or as Bonnie and Clyde,
Here on this whirling planet, or on the Other Side,

The Greatest Blessing from Above,
Is to Love, and to be Loved

Forever, unconditionally
Then and now, eternally,

Shining brightly as the sun,
Merging as the Cosmic One.

By far, God's Greatest Glory,
Available to lowly mortals,

Is The Cosmic Love Story,
It opens Heaven's Portals.

Ed Close, 2018

PART X: CYCLES OF CONSCIOUSNESS AND TIME

Prophets of all lands and ages have succeeded in their God-quest. Entering a state of true illumination, nirbikalpa samadhi, these saints have realized the Supreme Reality behind all names and forms. Their wisdom and spiritual counsel have become the scriptures of the world....
My gurudeva, Jnanavatar Swami Sri Yukteswar (1855-1936), of Serampore, was eminently fitted to discern the underlying unity between the scriptures of Christianity and the Sanatan Dharma.
- *Premavatar* Paramahansa Yogananda, 1949

This section of this book is dedicated to the memory of Swami Sri Yukteswar Giri, guru of Paramahansa Yogananda, the founder of Self-Realization Fellowship (Yogoda Sat-Sanga Society of India), and who is, therefore, my param-guru.

Sri Yukteswar Giri, author of the book *The Holy Science*, was born May 10, 1855 in Serampore India. He left the physical body March 21, 1936, in Puri, and I entered consciously into the fetus my mother was carrying on that same date, March 21, 1936, in Pilot Knob Missouri, on the opposite side of planet earth. But let me be very clear: I am *not* saying that I am the reincarnation of Swami Sri Yukteswar Giri. My param-guru, Jnanavatar Yukteswarji ascended to the astral planet *Hiranyaloka*, when he exited his body on March 21, 1936, to serve God as a teacher of souls who are blessed to be able to go there after attaining *nirbikalpa* samadhi. I came to Planet Earth as a small spark of that consciousness. Remember, in spirit, all consciousness is One. I have written about my memory of the experience of my most recent descent into this limited existence, and it was published in the November-December 2008 issue of *Telicom*, the Journal of the International Society for Philosophical Enquiry (Vol. XXI, No. 6, page 72, Autobiographical Sketches).

To paraphrase Albert Einstein: Time is an illusion, but it is a frustratingly persistent illusion as long as you are identified with, and attached to a physical body. In TDVP, we discover that time is three-dimensional. When I presented this concept to Stephen Hawking in an early manuscript of my second published book, *Infinite Continuity*, he responded that he couldn't imagine 3-D time. I had thought that because he was not entirely present in his body, he might have experienced three-dimensional time; but evidently, he had not. Now that he is free of his burdensome physical body, perhaps he can. It depends on his state of spiritual virtue. Time is an illusion because we only see one unidirectional dimension of it with the limitations of our physical senses. One who has ascended sees the whole picture. Human history, as we perceive it in the five-dimensional space-time-consciousness domain of our earth-bound existence, is restricted to a brief, transient bubble, a tiny fragment of nine-dimensional reality.

Recent History: the Past 12,500 Years

There is considerable evidence that sentient beings have lived in the cosmos, in this galaxy and on this planet for a very long time. But let's focus on some important events of the past 12,500 years, related to the rise and fall of civilizations on this planet in ways that fit into specifically-explainable logical patterns analogous to the patterns of the cube. 12,000 years is barely a blink (less than one one-hundred-thousandth of the estimated 13.8 billion-year estimated age of the big-bang universe). Notice that all objects that remain stable for thousands of years rotate (spin), and revolve in orbits around other, larger objects and/or around common centers (centroids) with other objects in a given group. For example, planets rotate on their axes, their moons revolve around them, and planets with their accompanying satellites revolve in elliptical orbits around stars like our sun. Furthermore, systems like our solar system, revolve around other larger systems and the center of the galaxy in which they exist.

Our solar system moves in its orbit within the Milky Way Galaxy at about 540,900 miles per hour. Not only that, galaxies rotate and revolve around other intergalactic centers as well. From quarks to quasars, the rule is periodic rotation and revolution, repeating over and over, like clockwork. We've shown in TDVP applications of the calculus of dimensional distinctions that mathematically, there are nine finite dimensions, three of space, three of time and three of consciousness. In this spinning, hyper-dimensional domain, all things, including physical development, mental and spiritual virtue are periodic and cyclic.

Spacetime and Time Cycles

As pointed out earlier (**Part IV, The Scientific Basis of Infinite 3-D Time**), The idea of something from nothing (creatio ex nihilo) is not supported by empirical evidence, and Einstein's general relativity shows that the measurement of spacetime is relative to the motion and proximity of the observer to massive objects. So time is different in different inertial systems, and in different parts of the universe. All available evidence supports the mathematical accuracy of describing dynamic change in cycles advancing through multi-dimensional spacetime, not absolute beginnings and ends. The way this works has been described by Swami Sri yukteshwar Giri, a man revered by many, including me, as *jnanavatar* (incarnation of wisdom).

In *The Holy Science*, written in India in 1894, Swami Sri Yukteswar Giri, an astronomer and life-long student of Vedic wisdom, calculated the period of the rise and fall of consciousness on this planet to be about 24,000 years, coinciding with the time it takes the sun to complete one revolution around its dual star, finishing one cycle, consisting of 12,000 years in an ascending arc, and 12,000 years in a descending arc, for a total of 24,000 years. (See **Figure Six**.) He says:

"*When the sun in its revolution round its dual comes to the place nearest to this grand center, the seat of Brahma (an event which takes place when the autumnal equinox comes to the first point of Aries), dharma, the mental virtue, becomes so much developed that man can easily comprehend all, even the mysteries of Spirit ... in a period of 24000 years, the sun completes the revolution around its dual and finishes one electric cycle consisting of 12000 years in an ascending arc and 12000 years in a descending arc.*"

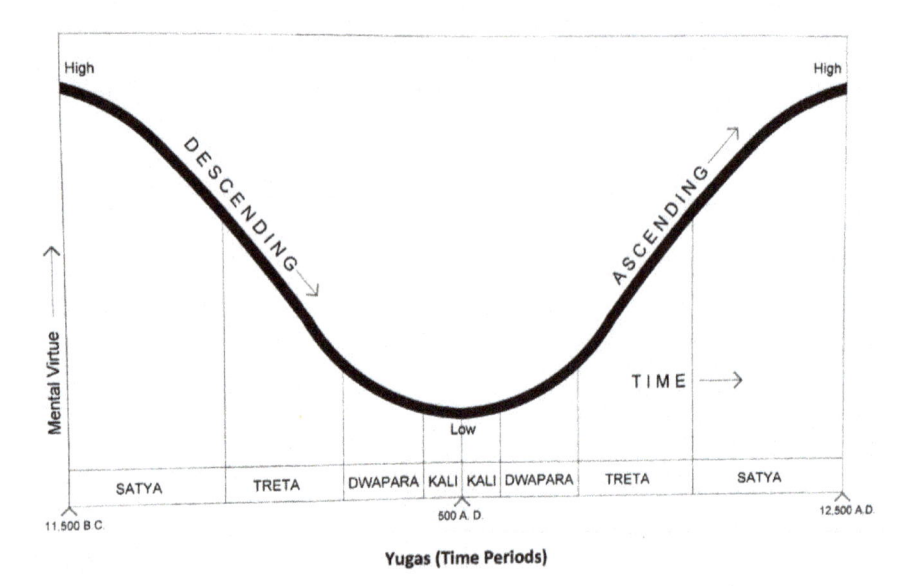

FIGURE SIX

In addition, he calculated that, at the time he was writing the Holy Science in 1894, we had advanced 194 years into the ascending Dwapara Yuga, the first period above the lowest point. Since the ascending Kali Yuga (the ascending half of the lowest Yuga, the period of nearly total absence of mental virtue, science and spirituality) is 1200 years, the nadir of development in this cycle was about 500 AD, very close to the year (553 AD) when Justinian usurped the power of the Pope and subverted the teachings of Jesus and the writings of Origen. We are now, in 2018, about 318 years into the ascending Dwapara Yuga. This means that the rise and fall of the mental virtue of sentient beings on this planet has occurred four times in the last 110,000 years, and we are 318 years into the fifth ascent.

Using Sri Yukteshwar's method, we can easily calculate the high and low points of mental virtue and spiritual understanding during the past 110,000 years. As we've seen, the last low point was around 500 AD. That means that the last high point occurred about 11,500 BC. From this, we can locate each of the low- and high-points of mental virtue in the last 110,000 years by moving back in 12,000-year and 24,000-year steps respectively. See Figure Six.

Is there any empirical evidence supporting the existence of a high level of human development more than 10,000 years ago to be found on planet Earth? Yes, there is: In Southern Turkey, a mound called Göbekli Tepe (Round-belly Hill) has been excavated, and the oldest layer has been reliably dated to have been used from 11,600 to 10,800 years ago. It included 10-ton stone pillars decorated with stylized bas-relief sculptures of animals, distorted human arms and hands and cryptic symbols. There are other archeological sites with stone structures and carvings that very likely cannot be duplicated today, even with our modern technology. However, most of them are dated by archeologists as built more recently than the stone pillars of Göbekli Tepe: The Egyptian Pyramids, 2,700 to 7,000 years ago, Stonehenge, about 5,000 years ago, Mayan ruins, 1,000 to 3,000 years ago, Puma Punku, Nasca Lines, etc., about 1,500 years ago.

At first glance, this evidence doesn't seem to fit into Sri Yukteswar's cycles of time. But this is because we have the erroneous idea that time is a uniform backdrop within which all events occur in a linear fashion. We are drawing the wrong conclusion because our ingrained idea about spacetime is false. It is based on the short-term appearance that time is linear and unidirectional, and that, along with space, time is everywhere uniform. But this completely ignores Einstein's general relativity, and a key concept he expressed in his final addition to his work on relativity. He saw space and time as extensions in the field of the substance of reality, with no existence of their own. When we combine this understanding with Sri Yukteswar's time cycles, things begin to make sense.

Even with modern science only a few hundred years old, we like to think that, if the mental virtue (scientific and spiritual understanding) of sentient beings was at an apex in 11,500 BC, then there should have been great cities with towering buildings, and wonderous machines all over the planet, yet there seems to be no evidence of that. Why? First, a lot may have been obliterated from the surface of the planet by erosion, plate tectonics, and other geologic processes, and even by human activity, in 12,000 years of descent. Second, 12,000 ascending years of mental virtue and spiritual advancement may not result in anything even vaguely similar to what we have now, near the low point, things that we have developed within just a few hundred years. It is entirely possible that conscious beings may evolve mentally and spiritually in 12,000 years to the point where buildings, machines, and even physical bodies would be completely unnecessary. Imagine a planet returned to its natural pristine condition!

According to general relativity, the spacetime of a mass-energy system is defined by the periodicity of the rotation and revolution of objects within the system. Our hours, days and years are defined by the movements of the objects within our solar system, and what we call space, is defined by the mass-energy-consciousness field relationships between the objects of the system. Time within another solar system, thousands of light years away, may be passing at a very different rate relative to ours, because the rotations, revolutions and the masses of the objects in that system are different than in ours, and because of their motion relative to us.

PART X: CYCLES OF CONSCIOUSNESS AND TIME 371

The relationship of the rise and fall of the mental virtue of consciousness to the most dominant periodic cycle of rotation in the physical system within which it exists, as stated by Sri Yukteswar is reasonable because, as discovered in the development of TDVP, the stability and thus specific rotational dynamics of physical system is produced by the presence of gimmel in the atoms. And those stable structures provide support for organic life, the vehicle of individualized consciousness. Because the mass-energy ratio and distribution is different in different systems, the development of mental virtue may proceed at different rates in different solar systems.

`The effects of one inertial system on another will diminish with distance, and at some point, it will not be just insignificant, it will be *zero*, because mass, energy and consciousness are quantized in physical reality, and the field effects of one solar system on another is also quantized. Like gravity, all energetic physical effects diminish with distance, so at some distance, it diminishes to one quantum of force, and with one more quantum of distance, it becomes zero. Thus, across intergalactic space, the spacetime dimensional domain of a star system in another galaxy may be vastly different than ours, and have no effect on ours. Thus, sentient life and its effects in solar systems in another galaxy may be totally undetectable for us until we find a way to travel to a point much closer to that galaxy. However, for star systems within our galaxy, interaction is possible, and development may proceed at different rates, depending on the total mass-energy dynamics of each system.

What about those sites with stone monuments, built with amazing engineering skill and precision, more than 1000, and less than 10,000 years ago, widespread around the world?

Were they really built by the ancestors of the people now living there who were presumably less advanced than their descendants, who had nothing but primitive tools? Here are three possible answers to this question:

1. Most of these sites were built at a time of much higher mental virtue than spiritual understanding in the last descending age than where we are now in the new ascending age (see **Figure Seven**), so they may have had memories and/or records of higher knowledge from the last age of highest development that was lost as the world passed through the lowest point about 500 AD, when the Great Libraries in Alexandria, Caesarea and elsewhere were burned.

2. Some of the sites may be far older than archeologists estimate, dating back to previous high points of development.

3. They may have been built by extraterrestrials from more advanced civilizations within our galaxy, as temporary bases for various reasons.

It is conceivable that all three of these answers may be correct for some of the sites. There is some evidence for the first two; is there any evidence of the third one? For extraterrestrials?

Let's look at Sri Yukteswar's time cycles.

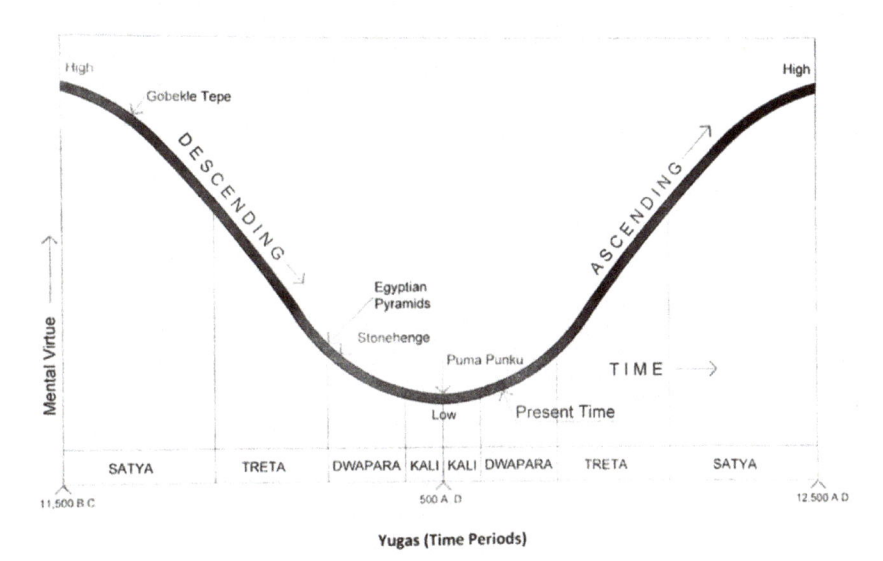

FIGURE SEVEN

Have Extraterrestrials visited Planet Earth?

There are a number of very reputable people, including astronauts Gordon Cooper and Edgar Mitchell, several past US Presidents, former Russian leader Mikhail Gorbachev, and Paul Hellyer, former Defense Minister of Canada, who have publicly stated their belief in the extraterrestrial origin of UFOs. Many scientists, including Stephen Hawking, believe the universe very likely contains intelligent life other than human beings, but caution that they may not be friendly. Some people, for example, Paul Hellyer, Erich von Däniken and Ancient Aliens theorists like David Childress and Giorgio Tsoukalos, believe ETs have been visiting Earth for thousands of years, have influenced human history and walk among us *incognito*, even today.

What do I think? I think the discovery of gimmel, evidence of consciousness in every stable atom, implies that organic life, capable of supporting consciousness, is intrinsic and inevitable in any stable portion of the universe. Therefore, I think that it is very likely that there are numerous sentient life forms in various stages of development in star systems throughout the universe. Whether or not they have been, or are currently visiting this planet, is another question, but it is probable because of the number of planetary systems that are known to exist in our galaxy. Indisputable evidence of such visitations is difficult to come by for a number of reasons. Most, if not all, of the massive monolithic stone artifacts in Turkey, Egypt, and South America, mentioned above may be explained by answers 1 and 2 in the previous section. While it is possible that some of them were built by extraterrestrials, proving that they were is difficult with the evidence on hand. The more recent sites, like Puma Punku are better arguments for extraterrestrial visitation.

Evidence sought as proof by believers in extraterrestrial visitations is usually in the form of 1) materials containing elements or compounds not found on this planet, 2) mummified or skeletal remains with DNA dissimilar to that of indigenous life forms, or 3) localized conditions or circumstances that cannot be explained any other way.

Extraterrestrial Materials?

Hundreds of tons of extraterrestrial material impact this planet every year in the form of cosmic dust and meteorites. So, finding materials with a mix of elements or compounds not found in geologic structures or manufactured on this planet, does not necessarily mean that it is part of an artifact constructed elsewhere in the universe and brought or sent here by space aliens. So it is difficult to make a conclusive argument for ET visitation based on the composition of unusual material objects of unknown origin.

Extraterrestrial DNA?

Are there people walking among us who are extraterrestrial aliens? Maybe. Do some of us, perhaps many, or most of us, have extraterrestrial DNA? This I think is more likely, and even probable. The popular single-origin mitochondrial Eve theory is in many ways analogous to the mathematical-singularity, big-bang theory of the origin of the physical universe, and it has the same problem as the unprovable big bang theory: It is logically untenable. The theory hypothesized one hominoid female, living in East Africa as the ancestor of all homo-sapiens, i.e., modern human beings. But, of course, that female, in order to give birth, had to be impregnated by a hominoid male, and he would have had a mother, etc., and all of these ancestors may or may not have come from the same location as the oldest found fossil to date. Some could have been extraterrestrial, or there could have been extraterrestrial interbreeding anywhere down the line, producing DNA that would be identified as part of the African haplogroup.

Haplogroups

Haplos means 'single' or 'simple' In Greek. Geneticists use the word 'haplogroup' to mean members of a group who, by virtue of similar DNA features, are descended from the same group or clan of ancestors. There are three major haplogroups of humans on planet Earth: Asian, African and Northern European. This cannot actually be proved definitively, because the percentage in a modern human's DNA from one ancestor who lived 100,000 years ago is far too small to measure with today's technology. If there are three different lines of DNA, one possibility is that the African haplogroup may have been impacted by two extraterrestrial migrations during the past, producing branches that became the Asian and Norse haplogroups.

This impact, however, may be obscured by on-planet migrations occurring after the ET impacts. If every human born has two parents, then the contribution is halved each generation. Half of your DNA came from each parent, one fourth from each grandparent, one eighth from each great-grandparent, and so forth. If the average generation (the time from being born to giving birth) is taken to be 20 years, in 100 years, i.e., five generations, each ancestor contributes 1 divided by $2^5 = 1/32 = 0.03125$, or 3.125%. In ten generations, the percentage is $(½)^{10} = 1/1024 = 0.00097656$, or less than one-tenth of one percent. 100,000 years is 5,000 generations, so the contribution a single ancestor is infinitesimal. Projections into the past beyond the seventh generation (where each ancestor contributes less that eight tenths of one percent) are highly speculative. Conclusion: proof of ancient-aliens genetic influence is problematic.

If we accept the vedic yugas time-cycle perspective, then sites with multi-ton stone artifacts that cannot be duplicated with our most modern technology today, may still have been constructed by human beings, but by people with knowledge superior to ours, if they were constructed more than 3,000 years ago, because the average mental virtue was more highly developed then than now. So there is no need to invoke aliens from other worlds to explain sites like Göbekli Tepe and the Egyptian Pyramids. But what about sites plausibly dated to times that are lower on the descending or ascending cycle than we are today? If they contain artifacts that are beyond our capabilities today, we may have evidence suggestive of extraterrestrial activity. Most artifacts confidently dated in the descending side of the cycle are relatively primitive, compared to more ancient artifacts like those of Ancient Egypt, and do fit fairly well into the Vedic view of cyclic mental virtue. There is, however, at least one exception.

Puma Punku

Puma Punku (Door of the Puma) is a complex of mega-ton stone artifacts, part of an archaeological site known as Tiahuanacu in Bolivia. Based on carbon dating of organic materials excavated from the foundational mound upon which the artifacts stand, archeologists date the complex to about 500 AD, and suggest that it must have been constructed by a stone-age civilization that existed prior to the Inca Empire, between 300 and 1000 AD.

The Puma Punku base was a terraced earthen mound originally held in place with intricately carved andesite blocks, each weighing several tens of tons. Andesite is a very hard igneous rock formed from hardened magma much like basalt, porphyry and granite. The stones were cut with such precision that they fit together to interlock with each other so perfectly that a razor blade cannot be inserted between them. The technical skill and engineering precision that it took to produce these stone blocks would be difficult, if not impossible to accomplish with power tools today. Some of the blocks exhibit machine-quality finished surfaces, with linear slots and holes drilled to perfection. The red sandstone slabs forming part of the terraces are estimated to weigh up to 131 metric tons (more than 144 US tons).

The Puma Punku complex is assumed to have been constructed by an agrarian people, ancestors of the present-day *Aymara* Indians, who had no written language and were ignorant of the existence of even simple tools like the wheel. Their civilization peaked during the lowest point of the Kali Yuga, around 500 AD. (**See Figure Seven.**) Given the possibility that extraterrestrial beings may have visited our planet, who is more likely to have produced these artifacts that we cannot duplicate today, pre-Colombian Indians, or extraterrestrials? Clearly the ETs!

Even this, however, is not irrefutable evidence of ET visitation, because of possibility #2: The precisely-carved stone blocks *may be far older than archeologists estimate, dating back to previous high points of development in the descending cycle*. The Aymara Indians may have simply found the stone blocks and slabs already at that site, and managed to use them to shore up and pave the mound, which they certainly could have built. Note that the archeologists were able to date the mound, *not* the stone slabs; but even moving the megalithic slabs weighing tens of tons would have been quite a feat for pre-Colombian Indians. Have sentient beings from other planets visited our planet? It is certainly possible, but as we have seen, if the Vedic view of cyclic time, which is consistent with physical law and the mathematics of TDVP, is correct, then there is no irrefutable evidence from the archeological sites discussed above, as claimed by vocal ancient alien theorists on television.

As noted above, there are reputable scientists, astronauts and public figures who believe UFOs are not of this world, and some people who claim that our government, and other governments, have retrieved crashed spacecrafts and alien bodies. But until the governments disclose this, or the aliens themselves announce their presence and begin to interact with us openly, we have no real proof. I, for one, am skeptical of the claims of ancient alien theorists, and people who tell stories of encounters or abductions, and even detailed interaction with many different alien species. Knowing that human imagination, self-deception and the desire for attention, fame and fortune are strong motivating factors, without proof, we should remain skeptical. It may be entertaining to go down some of the many rabbit holes offered by some theorists, writers and convention lecturers, but be careful, you may forfeit your sanity in the process.

The Way to Enlightenment

Do you have to wait 5,682 years for the beginning of the Sat-Yuga, to begin to attain personal enlightenment? No. Sri Yukteswar points out that, in any time period, an individual soul may go through the Kali, Dwapara, Treta, and Sat states quickly with the help of one of the Spiritual Preceptors. In the words of Shankaracharya, an 8th century AD incarnation of Krishna:

> *Life is always unsafe and unstable,*
> *Like a drop of water on a lotus leaf,*
> *The company of a divine personage,*
> *Even for a moment, can save us.*

My beloved Jacqui, a devout Christian who attained enlightenment by practicing Kriya Yoga, embodied the path of a modern American spiritual seeker, and she recommended this approach to other Christians, as you see in her writings above. The point of Sri Yukteswar's Holy Science was that the teachings of Jesus and Krishna were one and the same, differing only in the forms of presentation, reflecting their different cultures and languages. In the Holy Science, Sri Yukteswar meticulously compares Sanskrit aphorisms from the Vedas with verses from the Holy Bible, showing that they are both expressions of the same truths. In his introduction he says:

"The purpose of this book is to show as clearly as possible that there is an essential unity in all religions; that there is no difference in the truths inculcated by the various faiths; that there is but one method by which the world, both external and internal, has evolved; and that there is but one Goal admitted by all scriptures. ...

"The method I have adopted in the book is first to enunciate a proposition in Sanskrit terms of the Orient sages, and then to explain it by reference to the holy scriptures of the West. In this way, I have tried my best to show that there is no real discrepancy, much less any real conflict, between the teachings of the East and the West. Written, as the book is under the inspiration of my param-Guru-Deva [*Babaji*], and in the Dwapara age of rapid development in all departments of knowledge, I hope that the significance of the book will not be missed by those for whom it is meant."

Of course I recommend The Holy Science to my readers. It is available from SRF and on Amazon for as little as ten dollars. It is less than 100 pages in length, but may require multiple readings and study for one to absorb its profound message. I will share only a little more of its wisdom here, to give a brief but accurate description of the way to enlightenment taught by the masters of all spiritual traditions throughout the ages. He says:

Following affectionately the holy precepts of these divine personages, man becomes able to direct all his [sense] organs inward to their common center – *Sensorium, Trikuti,* or *Sushumnadwar,* the door of the interior - where he comprehends the voice, like a peculiar knocking sound, the word Amen (*Aum*), and the Godsent luminous body of Radha, symbolized as the forerunner, John the Baptist in the Bible.

Vide Rev. III:14,20; and John I:6,8,23:

"These things saith the Amen, the faithful and true witness, the beginning of the creation of God,"

"Behold, I stand at the door and knock; if any man hear my voice, and open the door, I will come in to him and will sup with him, and he with me."

"There was a man sent from God, whose name was John."

"He was not that Light, but was sent to bear witness of that Light."

"He said, I am the voice of one crying in the wilderness:

Make straight the way of the Lord!"

From the peculiar nature of this sound, issuing as it does like a stream from a higher region and losing itself in the lower creation, it is figuratively styled by various sects of people by the names of different rivers which they consider as sacred; e.g., the Ganges and Jumna by the Hindus, Buddhists and Jains, and the Jordan by

Christians and Jews.

Through this luminous body, man, believing in the existence of the true Light -the Life of this universe- becomes baptized or absorbed in the holy stream of the sound. This baptism is, so to speak, the second birth of man and is called Bhakti Yoga, without which man can never comprehend the real internal world, the kingdom of God.

Vide John I, 9, and III, 3:

"That was the true Light which lighteth every man that cometh into the world."

"Verily, verily, I say unto thee, Except a man be born again, he cannot see the kingdom of God."

PART XI: SUMMARY AND CONCLUSION

The Cosmic Love Story

Love is Eternal; as old as the sun and moon, as new as tomorrow; as obvious as a kiss, as hidden as your DNA; as simple as a puppy, as complex as an Einstein's thoughts. Love is the beginning and the end; as plain as a clear blue sky, as colorfully variable as a scrambled Rubik's cube. Love is the lingering glow of things now gone, the promise of things to come. Love is the ultimate puzzle, an enigma deeper than solving the cube. Love is the irresistible force that expands the universe in seven directions: up, down, right, left, ahead, behind, and inward, and it is the inward expansion that is the most important, because it leads to the kingdom of God!

From the day that beautiful young girl said "You're Ed Close!" to the day she said *"I want your face to be the last thing I see, before I see the face of God."*, my life has been filled with unconditional love. Together, we experienced the vicissitudes of life: the agonies of defeat and failure, and the joy of ultimate success. By the grace of God, with the loving assistance of the divine personages who are our Spiritual Preceptors, the torturous road of pain and suffering is now behind us; the sacred mathematical dimensionometry of our alignment with the Infinite is now complete. The puzzle has been solved. And my job now is to tell our story to all who have ears to hear; to all who can benefit from hearing it.

Physics, Spirituality and the Cube

All things are subtly connected. By defining the basic unit of quantum equivalence as the naturalized unit of measurement of the rest mass, energy and volume of the electron, and incorporating the relativistic effects of the spin of elementary subatomic objects, we developed the basic Triadic Rotational Unit of Equivalence (TRUE), and the cube is transformed into an effective model of physical reality at three vastly different scales of measurement and observation: the quantum-, macro- and galactic-levels. This allowed us to incorporate the action of consciousness, the drawing of a distinction, into a primary triadic mathematical system of logic and use the TRUE as the basic unit of the system which I call the Calculus of Dimensional Distinctions (CoDD).

Because the basic unit of the CoDD is based on the physical characteristics of the electron, and empirical data from the Large Hadron Collider (LHC), the forms of the primary arithmetic (number theory) and the primary algebra of the CoDD are directly related to the most basic measurements of physical reality, and thus reflect the logical structure of reality. The most useful feature of this unique approach is that it provides a logical mathematical method capable of determining which distinctions are real, which are conceptual, and which are extra-dimensional. This makes it a powerful tool for testing hypotheses about the relationship between consciousness and objective reality. Rather than the binary 'true' or 'false', logically 'valid' or 'invalid', hypotheses expressed in the language of the CoDD can be demonstrated to be triadic: i.e., either real, conceptual or imaginary, relative to the axiomatic system within which it is described.

Using this approach, we discovered that the existence of a third form of reality, in addition to mass and energy, is necessary for physical stability. This third form is mathematically and dimensionometrically necessary in order for stable physical structures to form and continue to exist for more than a few nanoseconds. It possesses no mass or energy, but contributes to the total angular momentum of high-energy spinning quanta, allowing quarks consisting of certain specific whole numbers of quantum equivalence units, to combine volumetrically to form the symmetrically stable subatomic objects we know as protons and neutrons.

All the atoms of the periodic table are built up of these stable objects, and life-supporting elements and compounds are found to have the highest ratios of this third form of reality, which Dr, Neppe and I call *gimmel*. The mathematical structure of quantized reality reflected in the cube conveys the logical patterns that allow the development of meaningful relationships between mass, energy, spacetime, and consciousness, suggesting that the quantized physical universe is designed by the pre-existing underlying logic of Primary Consciousness in such a way as to support the formation of organic life forms manifesting consciousness.

The CoDD mathematical relationships between the various aspects of physicality and consciousness comprise the laws of nature. With the inclusion of the functioning of consciousness in the equations of the CoDD, the laws of form are expanded to include the laws that govern the experiences of individual conscious entities. From this perspective, the purpose of the physical universe is to support conscious life and provide an arena for growth and development from rudimentary self-awareness to the awareness of relationship, and the eventual spiritual realization of the cosmic Self.

The Message of TDVP, the Cube and the Science of the Future

The message of the Triadic Dimensional Vortical Paradigm (TDVP) is that the essence of reality manifests triadically in the physical universe as mass, energy and consciousness, interacting in a nine-dimensional space-time-consciousness domain, the workings of which are comprehensible to human beings because the consciousness, bodies and essence of individual life forms are composed of the same elementary quantum equivalence units of substance as the quantum, macro and galactic aspects of the universe, and the mental and spiritual nature of human beings reflects, albeit, sometimes imperfectly, the logic of Primary Consciousness.

The cube consists of three sets of three mutually orthogonal planes of nine cubes each that are free to rotate around a common center. The equivalent sub-cubes are analogous to Triadic Rotational Units of Equivalence (TRUE quantum equivalence units), and the three interacting domains of the cube can be used to simulate the way the space, time and consciousness domains interact at the quantum, macro and cosmic levels. This leads to the realization that **the conscious field that holds everything together in functional form, and the fabric within which it is all suspended is nothing other than the attraction of Divine Love.**

The Calculus of Distinctions is a primary calculus from which all mathematical forms and operations are derivable. The Calculus of *Dimensional* Distinctions, with the TRUE quantum equivalence unit as its basic unit of observation and measurement, replaces the calculus of Newton and provides a means to properly describe physical, mental and spiritual phenomena.

By incorporating the actions of primary and individualized consciousness into a consistent logical system, application of the CoDD reveals the necessity of the existence of *gimmel*, the non-physical third form of the essence of reality, that is not measurable as mass or energy. **Neither matter nor energy, gimmel is the non-severable link with the conscious fabric of reality, the infinite field of Divine Love.**

TDVP, *the science of the future*, integrates classical natural science, relativity and quantum physics; Eastern, Middle-Eastern and Western spiritual philosophy; and the physical sciences; psychology, medical science and the perennial philosophy of real mysticism. God's infinite love, finitized in the form of individual love for mother, father, wife or husband, children, other human beings, and all living beings, is what holds reality together, and drives the spiritual evolution of consciousness in the physical cosmos.

The message of this book is simple: Like the cube, life is a puzzle. Sometimes difficult, sometimes easy, with most of the inner workings hidden. As long as your life and consciousness are in disarray, like a scrambled cube, existence is filled with misery, pain and suffering. But, by becoming focused and aligned with the physical, mental and spiritual laws of the universe, you come to realize that you are eternally blessed. Through the gift of consciousness and life, the ultimate experience is available to everyone. We are, indeed, as Jesus said, sons and daughters of the most-high. In the end, **Alignment with GOD is Simple and Easy. Everything else is complex and difficult, leading to virtually endless complications. So, our advice is: Keep the goal of Cosmic Consciousness in mind, open your heart to the infinite attractive potential of Divine Love, and enjoy the journey!**

Ode to Joy Now Gone

Brisk winter day; the sun is warm; the ground is frozen hard.
The puppy, Maximus, and I go out and walk around the yard.
He's grown nearly thrice in size since the day my sweet love died;
And I'm blessed to have this companion now, trotting by my side.

We visit the shade gardens that Jacqui and I planted - oh, so many years ago,
Now covered in dead leaves and weeds, bulbs waiting yet for spring, I know.
Looking up at the windows of the house where we shared so much love,
My heart aches for the Joy now gone, as Jacqui looks at me from above.

This was our haven, amid towering oak and hickory trees,
High above the Mighty Mississippi's flow,
On the Trail of Tears, where the many hapless Cherokees
Fell and died, one hundred eighty years ago.

Perhaps it was no accident that Jacqui, my love,
Who was of the Deer Clan of the Cherokee,
Would leave this life in this cold Missouri grove,
Far from her ancestral home in Tennessee.

Sheltering from the icy wind, to the sunny side of a stalwart tree,
Within, I turned a corner then; I looked at Max, he looked at me.
I had moved beyond the loss of the life we had those forty years,
I heard her whisper tenderly: "It's time to leave the trail of tears".

ERC January 27, 2019

NOTES

Jacquelyn A Close, RA, CCI has helped millions of people world-wide to live better, happier, healthier lives, by her tireless promotion of the idea that we should all take personal responsibility for our physical, mental and spiritual well-being. She has helped to educate thousands, if not millions, about the importance of the use of natural alternatives to allopathic and pharmaceutical health care, and the use of such natural, patient-friendly modalities in conjunction with, or as complimentary to, surgery and allopathic medicine if and when such extreme measures are necessary to save or the enhance the quality of life. She held many weekly and monthly on-line webinars about the use of essential oils, and travelled around the country and internationally with her husband, Dr. Ed Close to spread the word.

She also stressed the importance of treating the body as the temple of the Spirit of God that dwells within each one of us. While she was a devoted Christian from childhood, and until her last breath, she also appreciated the common core of all true religions. Instead of focusing on the differences between cultures and their organized religions, she focused on reverence for God, and the worship of the Infinite Intelligence behind all creation, regardless of the form it took. She was famous for her eagerness to help others, her generosity, and her bright, and cheerful personality. She lifted the mood wherever she went, and her ever-ready smile and laughter were infectious. It was hard not to smile in the glow of her presence.

Dr. Edward R. Close, PhD, PE has been recognized as a polymath: mathematician, theoretical physicist, cosmologist, hydrogeologist, and environmental engineer. A teacher of mathematics, science, languages and meditation techniques; a practicing mystic for many years, he has received awards and accolades from government agencies, universities, and high IQ societies.

From the Administration of the International Society for Philosophical Enquiry:

"We are pleased to announce that Dr Robert Campbell, ISPE Advancements Officer, has certified that Dr. Edward Roy Close has qualified for advancement to Diplomate, the highest earned rank in our Society.

The Charter (Article 3, Section 3f) requires a Diplomate to be a "person distinguished among colleagues by having demonstrated exemplary performance and achievement within the Society, as described by the Charter, and recognized for accomplishments outside the Society that may be construed as beneficial to humanity in general."

Dr Edward Close PhD, PE, SFSPE certainly markedly fulfills the Diplomate requirements. He is one of ISPE's most esteemed members, having achieved an international reputation as a Physicist, Mathematician, Cosmologist, Environmental Engineer and Planner, and international consultant on both mold remediation and on essential oils and health.

"Dr Close so easily fulfilled all the Diplomate requirements per the Charter that, for example, during this 11/2008 to 6/2014 period alone, he can list fifty highly meritorious achievements (independent of his five-year work with Dr Neppe). This is far more than the current Charter requirement for Diplomate which requires only sixteen significant and substantial contributions."

The ISPE President goes on to list more than 30 documented awards, accomplishments and achievements, including books and papers published, the creation of a major paradigm shift, and the solution of several long-standing problems in mathematics and physics.

Dr. Vernon Neppe, MD, PhD, also one of the rare Diplomates of the ISPE, has achieved an outstanding international reputation in several disciplines, including: Neuropsychiatry, Psychopharmacology, Behavioral Neurology, Forensic Psychiatry, and Neuropsychiatry. He has been honored as one of the ten top doctors in America. He is active in Consciousness Research, Phenomenology, Epileptology and Neuroscience, in which his contributions have been pioneering and he is also an internationally known Professional Speaker.

Dr. Neppe has pioneered in identifying the links of brain function and subjective experience. He is the world authority on déjà vu phenomena and has developed Phenomenology in the Neuroscience and Consciousness contexts. The author of nine books, and two plays (www.brainvoyage.com), he has published more than 400 publications. He has lectured in twelve countries, chaired international symposia, and worked internationally with the media and led the 1st USA and International Delegation in Neuropsychiatry and Psychopharmacology.

Originally from South Africa, where he began his education at the University of the Witwatersrand in Johannesburg, he earned a Fellowship at Cornell University, and established the first Division of Neuropsychiatry in a USA Psychiatry Department at the University of Washington, and founded the Pacific Neuropsychiatric Institute (PNI), which he directs (www.pni.org). He is also an Adjunct Full Professor in the Dept of Neurology and Psychiatry at St. Louis University, St. Louis, MO. He was the first USA based MD to be elected a Fellow of the Royal Society of South Africa, is a Distinguished Fellow of the American Psychiatric Association, recipient of the Marius Valkhoff prize, one of the rare Diplomates of the International Society for Philosophical Enquiry (ISPE) (www.thethousand.com), and Executive Director of the Exceptional Creative Achievement Organization (5eca.com). tdvp@pni.org.

REFERENCES

Nourse, James G, *The Simple Solution to Rubik's Cube*, Bantam Books, 1981

Singmaster, David, *The Utility of Recreational Mathematics*, Cambridge University Press, 1994.

Wheeler, JA, *At Home in the Universe*, American Institute of Physics, AIP Press, Stanbury, NY, 1994

Bucke, Richard M, *Cosmic Consciousness*, Penguin Books, New York, 1901

Close, ER, *The Book of Atma*, Libra Publishers, New York, 1977.

Close, ER, *Infinite Continuity*, Paradigm Press, Los Angeles, 1990.

Close, ER. *Transcendental Physics*, iUniverse.com Inc. Lincoln Nebraska, 1997

Close, Edward R, *Big Creek History, Folklore and Trail Guide*, Paradigm Press, Jackson MO, 2003

Neppe VM, Close ER, *Reality begins with consciousness: a paradigm shift that works (5th Edition)* Fifth Edition. Seattle, 2014: Brainvoyage.com.

Yukteswar Giri, Swami Sri, The Holy Science, Self-Realization Fellowship, Yogoda Sat-Sanga Society, India, 1957

Schrödinger, Erwin, What is Life? Cambridge University Press, 1992.

Neppe VM, Close ER, The fourteenth conundrum: Applying the proportions of gimmel to Triadic Rotational Units of Equivalence compared to the proportions of dark matter plus dark energy: Speculations in cosmology._IQNexus Journal 7: 2; 72-73, 2015.

Close ER, Neppe VM, Speculations on the "God matrix": The third form of reality (gimmel) and the refutation of materialism and on gluons._World Institute for Scientific Exploration (WISE) Journal 4: 4; 3-30, 2015.

Neppe VM, Close ER: Speculations about gimmel Part 5._World Institute for Scientific Exploration (WISE) Journal 4: 4; 21-26, 2015.

Close ER, Neppe VM: Preliminary ideas on gimmel that need confirmation. Part 4._World Institute for Scientific Exploration (WISE) Journal 4: 4; 18-20, 2015.

Neppe VM, Close ER: Key ideas: the third substance, gimmel and the God matrix. Part 1._World Institute for Scientific Exploration (WISE) Journal 4: 4; 3-4, 2015.

Close ER, Neppe VM: Summary and conclusion gimmel, TRUE and the structure of reality (Part 20). IQNexus Journal 7: 4; 112-114, 2015.

Close ER, Neppe VM: More questions answered on the elements, TRUE and gimmel (Part 17)._IQNexus Journal 7: 4; 82-102, 2015.

Close ER, Neppe VM: Hydrogen and the elements of the periodic table: applying gimmel (Part 13). *IQNexus Journal* 7: 4; 66-69, 2015.

Close ER, Neppe VM: The TRUE unit: triadic rotational units of equivalence (TRUE) and the third form of reality: gimmel; applying the conveyance equation (Part 12). *IQNexus Journal* 7: 4; 55-65, 2015.

Close ER, Neppe VM: Empirical exploration of the third substance, gimmel in particle physics (Part 10). *IQNexus Journal* 7: 4; 45-47, 2015.

Close ER, Neppe VM: Introductory summary perspective on TRUE and gimmel (Part 1) in Putting consciousness into the equations of science: the third form of reality (gimmel) and the "TRUE" units (Triadic Rotational Units of Equivalence) of quantum measurement *IQNexus Journal* 7: 4; 815, 2015.

Close ER, Neppe VM: Putting consciousness into the equations of science: the third form of reality (gimmel) and the "TRUE" units (Triadic Rotational Units of Equivalence) of quantum measurement *IQNexus Journal* 7: 4; 7-119, 2015.

Neppe VM, Close ER: The gimmel pairing: Consciousness and energy and life (Part 13D). *IQNexus Journal* 7: 3; 122-126, 2015.

Anonymous. Planck mission full results confirm canonical cosmology model. Dark matter, dark energy, https://darkmatterdarkenergy.com/2015/03/07/planck-mission-full-resultsconfirm-canonical-cosmology-model/. 2015.

Collaborators P: Planck Publications: Planck 2015 Results European Space Agency. *Astro-ph CO* Febr, 2015.

Cowen R, Castelvecchi D: European probe shoots down dark-matter claims. *Nature Physics* Doi:10.1038/nature.2014.16462, 2014.

Collaborators P: Planck 2013 results. XVI. Cosmological parameters. *Astro-ph.CO* arXiv:1303.5076, 2013. Collaborators P: Planck 2013 results. I. Overview of products and scientific results. *Astro-ph.CO* ArXiv:1303.5062, 2013.

Neppe VM, Close ER: A nutshell key perspective on the Neppe-Close "Triadic Dimensional Distinction Vortical Paradigm" (TDVP). *IQNexus Journal* 7: 3; 7-77, 2016.

Anonymous. Abundance in the universe of the elements. from http://periodictable.com/Properties/A/UniverseAbundance.html. 2016.

Wikipedia. Abundance of the chemical elements. Retrieved August 2016, from https://en.wikipedia.org/wiki/Abundance_of_the_chemical_elements. 2016.

Anonymous: June 13, 2011. Dark matter. 2011, from http://en.wikipedia.org/wiki/Dark_matter. 2011.

Zimmerman Jones A. Hawking and Hertog: String Theory can explain dark energy. June 20, 2006, from http://physics.about.com/b/2006/06/20/hawking-hertog-string-theory-can-explain-darkenergy.htm. 2006.

Close ER, Neppe VM: The Calculus of Distinctions: A workable mathematicologic model across dimensions and consciousness._Dynamic International Journal of Exceptional Creative Achievement* 1210: 1210; 2387 -2397, 2012.

Close ER, Neppe VM: Mathematical and theoretical physics feasibility demonstration of the finite nine dimensional vortical model in fermions._Dynamic International Journal of Exceptional Creative Achievement* 1301: 1301; 1-55, 2013.

Close ER, Neppe VM: Fifteen mysteries of 9 dimensions: on Triadic Rotational Units of Equivalence and new directions, Part III._Neuroquantology* 13: 4; 439-447, 2015.

Bottomley J, Baez J: Why are there eight gluons and not nine?, pp http://math.ucr.edu/home/baez/physics/ParticleAndNuclear/gluons.html 1996.

Anonymous: May 13, 2011. Gluon. 2011, from http://en.wikipedia.org/wiki/Gluon. 2011.

Feynman RP: *Electrons and their interactions. QED: The strange theory of light and matter*. Princeton, New Jersey: Princeton University Press., 1985.

Feynman RP (ed.). *The Feynman lectures on physics*. USA, Addison-Wesley, 1965.

Neppe VM, Close ER: *Relative non-locality - key features in consciousness research* (seven-part series) *Journal of Consciousness Exploration and Research* 6: 2; 90-139, 2015.

Close ER, Close JA: *Nature's Mold Rx, the Non-Toxic Solution to Toxic Mold*, EJC Publications, Jackson. Missouri, 2007.

Stewart, David, *The Chemistry of Essential Oils Made Simple, God's Love Manifest in Molecules* CARE Publications, Marble Hill, Missouri, 2005

Stewart, David, *Healing Oils of the Bible*, CARE Publications, Marble Hill, Missouri, 2003, Thirteenth printing, 2016.

Golden, Chuck, *Building and Living the American Dream*, Scotts Printing, Rolla, Missouri, 2007

INDEX

A

AAPS (Academy for the Advancement of Postmaterialist Sciences), 300, 304

Amman Jordan, 171, 174

Ancient Egypt, 166, 272, 377

Anathemas, the, 92, 93, 186, 187

Arabic, 168, 205, 282, 283, 285, 288

Aramaic, 44, 195, 202, 274

ASCSI (Academy for Spiritual and Consciousness Studies, Inc.), 300

Australia, 143, 161, 236, 299, 302

Abydos Egypt, 165

B

Babaji, Mahavatar, 133, 277, 286, 362, 380

Bachemin, Beverly, 350

Baltimore, Maryland, 145

Bible, the, 140, 142, 151, 173, 183, 184, 195, 201- 206, 220. 223, 251, 287, 324, 357, 380, 381

Big-Bang Theory, 65, 70, 182, 365, 375

Big Creek, 130, 158, 159, 393

Bobbi, 132, 135, 138, 140, 138-140, 148, 268, 269

Bohr, Niels, 50

Bonaparte, Napoleon, 216

Brisbane Australia, 302

Brown, G. Spencer, 63,

Bucke, Dr. R, 82

C

California, 131-133, 140, 142, 149, 153, 155, 268, 271, 278, 281

Cantor, Georg, 247

Cape Girardeau Missouri, 155, 159, 341, 343

Cayce, Edgar, 218, 227

Central Methodist College, 128

Centro, El Centro de Karma Yoga, 145, 146

Caesarea, 196-198, 372

Catholic Church, the, 183-185, 189-195, 202, 203, 223-225

Christ Consciousness, 198, 206, 257, 281, 317, 323

Christian doctrine, 93, 187, 189

Christianity, 93, 182, 183, 189, 194, 195, 201, 203. 242. 364

Cicero, 220

Close, Edward R (Ed), i-iv, 253, 264, 266, 270, 335, 350, 354, 355, 363, 383, 389, 390, 393-398

Close, Jacquelyn A (Jacqui), i, iii, 124, 148-155, 159-162, 171, 172, 235-237, 248-251, 257, 260, 267, 270-272, 276, 278, 280, 283, 386-392. 394, 398-305, 308, 320, 358-362, 379, 388. 389

Close, Joshua, 149, 150, 155, 156, 283. 286, 287, 304, 305, 319, 334, 341-343, 353

Congresso de la Sciencia y Spiritualidad, Puebla Mexico, 2013, 303

Consciousness, ii-iv, vii, vii, x, xiii-xvii, 33-36, 43, 62, 66, 77-80, 82-90, 93-96, 107-112 , 115, 122, 126, 134, 136, 137, 141, 143, 152, 153, 164, 168, 172, 177-182, 199, 206, 208, 209, 211, 224,

226, 229-231, 234, 236, 238-244, 246-248, 251-253, 257, 272, 276, 279-281, 299-302, 307, 321, 322, 336, 353, 364-367, 369, 371, 373-375, 384-387

Cosmos, 131, 135, 139, 143, 145, 149, 151-153, 155-157, 229, 283, 293, 298, 300, 417, 437

Cosmic Consciousness, xiv, xv, 82, 83, 88, 108, 115, 122, 177, 178, 181, 206, 224, 238, 240, 243, 244, 246, 272, 281, 387

Current River National Scenic Waterway, 129, 156, 158

D

Dali, Salvador, 217, 227

Delayed-Choice experiment, 44

Dimensionometry, xx, 85, 247, 383

Dimension, 63, 64, 94, 99, 365

Dimensional Domain, xv, 56, 64, 83-87, 98, 110, 247, 366, 371

Double-Slit experiment, 44

Duke University, 128, 300

E

Egypt, 163, 166, 176, 234, 272, 274, 276, 283, 302, 374, 377

Einstein, Albert, 64, 119, 120, 180, 189, 365

Electron, 37, 43, 47, 48, 51, 52, 54-61, 66, 67, 70-72, 74-76, 84, 85, 87, 152, 182, 253, 384

Emerson, Ralph Waldo, 215

Empedokles, 184, 274

Enlightenment, 82, 108, 111, 115, 116, 121, 131, 181, 199, 200, 227, 239, 240, 248, 272, 379, 380

EPR Paradox, 44, 50

Essential Oils, 143, 154, 160, 161, 236, 290-299, 302, 304, 311, 314, 328, 330, 346, 389, 390

Euclid, 97, 121

F

Fermat, Pierre, 121, 276

Fermat's Last Theorem, 35, 39, 42, 43, 58, 59, 64, 121, 163

Fermions, 48, 88, 397

Ford, Henry, 214

Ford, Kent, 102,

Fort Pierce, Florida, 233

Franklin, Benjamin, 213

French Revolution, 218

G

German (language), 114, 123, 194

Germany, 189, 190, 192, 193, 196, 233, 234

Giza, Egypt, 165, 166, 173, 196

Golden, Charles (Chuck), 156, 157, 398

Golden, Kay, 156, 157

Golden's Health Club, 157

Golden Hills, 156, 157

Golden Colorado. 278

Gödel, Kurt (incompleteness theorems), 24, 31, 73. 121

Greek 184, 195, 197, 202, 223, 274, 282, 375

H

Harrison, George, 218

Hawking, Stephen, 96, 218, 365, 373

Hebrew, 184, 195, 197, 202, 222

Heisenberg Uncertainty Principle, 92

Hijazi Mountains, 283

Hinduism, 243

Hitler, Adolph, 191-195, 216

Holy Roman Empire, 93, 189, 192, 193, 199

Holy Science, The Holy Science, 132, 279, 364, 367 368, 380, 393

Huxley, Thomas H, 137, 216

I

Incompleteness Theorem, See Gödel, Kurt

India, 149, 182, 210, 234, 243, 276, 364, 367

Indian (American Aborigines), 130, 378

Inertia, 44, 46-48, 66-69, 72, 73, 74

Infinity, 240, 247, 249

International Society for Philosophical Enquiry (ISPE), 162, 299, 302, 365, 390-392

Intrinsic spin, xvii

Intuition, ix-xii, 8, 20, 22-25, 77, 78, 81, 109, 110, 165, 173

Israel, 176, 357

Islam, 93, 188, 205

J

Jainism, 212, 381

James, William, 200, 208, 215

Jeddah, Saudi Arabia, 149-151, 155, 282, 283, 286, 287

Jebel Al-Radwah, 285

Jesus, 92, 118, 134, 138, 174, 175, 182=184, 186, 87, 194, 196, 198, 202, 204, 206, 215, 220-223, 275, 281, 354

John the Baptist, 174, 221-223, 380

Johns Hopkins University, 145

Jordan, 165, 171, 302

Jordan River, 174-176, 197, 222, 223

Josephus, 220

Judea, 92, 183, 215, 275

Jung, Carl, 219

Justinian I, Emperor, 92, 185, 186, 188, 189, 191, 192, 206

K

Klein, Oskar, xviii, 64

Kaluza, Theodor, xviii, 64

Koran, the Holy, 200, 205, 283

Kriya Yoga, 112, 134, 141, 268, 355, 379

L

Lawrence. Lee, 352

Laws of Form, 385

Lebanon, Tennessee, 257

Lebanon (Country), 284

Lennon, John. 218

LHC, Large Hadron Collider, xix, 47, 75, 384

London, Jack, 219

Lorentz contraction, 37

Luxor, Egypt, 165

M

McLaine, Shirley, 217

Mass, 36-38, 40, 44, 45, 47-53, 58, 59, 61, 62, 64-70, 72-76, 100-104

Mass, Conservation of, 36, 49, 51, 73, 91, 96, 178, 180, 228, 229

Mass of the electron, 37, 66, 67, 72, 74

Mass of the Neutron, 69, 73-75

Mass of the Proton, 48, 58, 68-74

Memory, xii, 12, 20, 22. 24, 77-80, 106, 176, 188, 205, 212, 217, 218, 232, 254, 266, 267, 273, 276, 321, 325-327, 333, 348, 351, 353, 354, 364

Missouri, 112, 114, 122, 123, 125-127, 129, 132, 135, 136, 139, 142, 150, 155-159, 163, 168, 233, 236, 253, 277, 292, 302-304, 335, 364, 388

Moore, Dr John, 269

Moses, 174

Mount Nebo, 174

Moya, Aury, 146, 147

Multi-dimensional, xix, 44, 64, 83, 94, 244, 366

N

Nashville Tennessee, 257, 258, 261, 264

Napoleon Hill, 123, 158

NDE (Near-Death Experience), 231, 301, 316, 319-322, 326, 328, 334, 348, 350, 352, 361

Neppe, Vernon M, i-iv, vii, xiii, 33, 35. 162-164, 298-300 302-304, 328, 350, 385, 391-396, 441

Newton, Sir Isaac, 41, 64, 98, 99, 386

Neutron, 50, 52, 56, 57, 69-71, 73-75, 84, 152

Nordström, Gunnar, xviii

Number Theory, 34, 43, 63, 384

O

OBE (Out-of-Body Experience), 143, 168, 231, 251

Origen (Oregenes), 92, 184-187, 195-198, 202, 204, 206, 220, 368

Orient Sages, 351, 380

Orientation, 2, 12, 15, 110

Origin event, 92

Ozarks, 129, 135, 150, 156

P

Parapsychology, 128, 300

Pasadena California, 148, 281, 286

Patton, General George S, 216

Pauli, Wolfgang, 64

Petra, Ancient City, 165, 166, 170-174, 234, 238

Pilot Knob Missouri, 364

Planck, Max, 33, 41, 46, 66, 100, 103, 104

Planck Probe, 100, 103, 104

Plato, 184, 274

Proton, 48-52, 54, 56-61, 68-71, 73-75, 84, 152, 180, 182

Puerto Rico, 115, 116, 145, 269

Puebla, Mexico, 303

Pyramid, the Bent, 165

Pyramid, the Great, 165-169, 171-173, 235

Pythagoras, 97, 184

Pythagorean Theorem 35, 38, 64, 121

Q

Quantum Physics, iii, iv, vii xiv, xviii, 35, 43, 45, 46, 76, 80, 173, 180, 299, 387

Quantum Equivalence Units, 33, 41, 66, 69, 75, 385, 386

R

Reincarnation, 145, 178, 181-185, 188, 189, 199-201, 205, 207-209, 211-222, 226-229, 364

Relativity, iv, vii, xiii, xiv, 8, 45, 76, 366, 368-370

Rhine, Dr. JB, 128, 129, 300

Rhine Research Center, 300

Rotation, 9, 19, 30, 34, 43, 63, 64, 67, 68, 74, 76, 79, 104, 268, 300, 366, 370

Revolution, 73, 218, 300, 366, 367, 370

Rubik's Cube, i-iv, vi, viii-xi, xv, xvii, 1, 3, 26, 31-34, 39, 43, 75, 109, 283, 383, 393

Rubin, Vera, 102

S

Sacramento, California, 140, 143, 268

Saudi Arabia, Kingdom of, 149, 150, 154, 155, 234, 281-283, 285, 287, 288

Schopenhauer, Arthur, 219

Schrödinger. Erwin, 394

Schwartz. Gary E, 350

Schweitzer, Albert, 214

Self-Realization Fellowship (SRF), 133, 134, 143, 217, 243, 268, 271, 278, 281, 299, 300, 303, 312, 313, 317, 351, 355, 364, 380

Spanish, 146, 190, 181, 217, 254, 282

Spin, xvii, 33, 34, 36, 37, 40-45, 48, 57, 64, 66-68, 71-76, 87, 88, 103, 104, 110, 134, 146, 168, 169, 172, 230, 253, 366, 384, 385

Stallone, Sylvester, 218, 227

Stewart, David M, Dr, 124, 125, 127, 128, 132-134, 142, 154, 156, 158, 160, 236, 268, 292, 300, 398

Stewart, Lee, 156, 160

Stevenson, Dr. Ian, 182, 209-213, 227, 350

Systems Analysis, 70, 82, 83, 84

T

Tampa, Florida, 148, 257, 259-261, 264, 267, 269, 271, 277

Tennessee, 257, 258, 261, 264

Texas County Missouri, 158

Thoreau, Henry David, 215

Trans-Arabia Pipeline, vi, 281

Transcendental Physics, vii, 35, 124, 299, 350, 393

Triadic Rotational Units of Equivalence (TRUE), 77, 386, 389, 394, 395

Triadic Dimensional Vortical Paradigm (TDVP), ii, vii, xvi, xix, xx, 34, 35, 40, 43, 46, 48, 56, 57, 66, 84, 102, 180, 247, 300, 302, 365, 366, 371, 378, 385-387, 392, 396

Twain, Mark, 116, 123

Tyndall, 140

Tyndall, George W, 234, 255,

U

Uncertainty Principle, 92

V

Voltaire, 215

Vortical, ii, xvi, xix, xx, 35, 44, 48, 88, 164, 386, 396, 397

Vortex, xx

W

Wheeler, John A, xvi, 393

Whitman, Walt, 216

Y

Yanbu Al Nakl, 284, 285

Yanbu Industrial City, vi, 149, 281-288

Young, D. Gary, 160, 161, 163, 236, 347

Young, Gary and Mary, 160-163

Young Living, 123, 124, 160-163, 165, 168, 169, 171, 172, 299, 302-304, 311, 314. 330, 331, 346, 347, 359, 360

Yogananda, Paramahansa, 133, 134, 135, 140, 217, 227, 243, 244, 248, 279, 307, 312, 317, 319, 352, 354, 364

Yukteswar, Jnanavatar, 132, 133, 244, 279, 280, 364, 367, 379, 371, 372, 379, 380. 393

Lightning Source UK Ltd.
Milton Keynes UK
UKHW020253051119
352876UK00008B/288/P